Modes of Parametric Statistical Inference

Modes of Parametric Statistical Inference

SEYMOUR GEISSER

Department of Statistics
University of Minnesota, Minneapolis

with the assistance of

WESLEY JOHNSON

Department of Statistics
University of California, Irvine

WILEY-
INTERSCIENCE

A JOHN WILEY & SONS, INC., PUBLICATION

Library of Congress Cataloging-in-Publication Data:

Geisser, Seymour.
    Modes of parametric statistical inference/Seymour Geisser with the assistance of Wesley Johnson.
        p. cm
    Includes bibliographical references and index.
    ISBN-13: 978-0-471-66726-1 (acid-free paper)
    ISBN-10: 0-471-66726-9 (acid-free paper)
    1. Probabilities. 2. Mathematical statistics. 3. Distribution (Probability theory)
    I. Johnson, Wesley O. II. Title.
    QA273.G35 2005
    519.5'4--dc22

                                                    200504135

Printed in the United States of America.

10 9 8 7 6 5 4 3 2 1

Contents

Foreword

In his Preface, Wes Johnson presents Seymour's biography and discusses his professional accomplishments. I hope my words will convey a more personal side of Seymour.

After Seymour's death in March 2004, I received numerous letters, calls, and visits from both current and past friends and colleagues of Seymour's. Because he was a very private person, Seymour hadn't told many people of his illness, so most were stunned and saddened to learn of his death. But they were eager to tell me about Seymour's role in their lives. I was comforted by their heartfelt words. It was rewarding to discover how much Seymour meant to so many others.

Seymour's students called him a great scholar and they wrote about the significant impact he had on their lives. They viewed him as a mentor and emphasized the strong encouragement he offered them, first as students at the University of Minnesota and then in their careers. They all mentioned the deep affection they felt for Seymour.

Seymour's colleagues, present and former, recognized and admired his intellectual curiosity. They viewed him as the resident expert in such diverse fields as philosophy, history, literature, art, chemistry, physics, politics and many more. His peers described him as a splendid colleague, free of arrogance despite his superlative achievements. They told me how much they would miss his company.

Seymour's great sense of humor was well-known and he was upbeat, fun to be with, and very kind. Everyone who contacted me had a wonderful Seymour story to share and I shall never forget them. We all miss Seymour's company, his wit, his intellect, his honesty, and his cheerfulness.

I view Seymour as "Everyman" for he was comfortable interacting with everyone. Our friends felt he could talk at any level on any subject, always challenging them to think. I know he thoroughly enjoyed those conversations.

Seymour's life away from the University and his profession was very full. He found great pleasure in gardening, travel, the study of animals and visits to wildlife refuges, theatre and film. He would quote Latin whenever he could, just for the fun of it.

The love Seymour had for his family was a very important part of his life. And with statistics on his side, he had four children—two girls and two boys. He was blessed with five grandchildren, including triplets. Just as Seymour's professional legacy will live on through his students and his work, his personal legacy will live on through his children and grandchildren.

When Seymour died, I lost my dictionary, my thesaurus, my encyclopedia. And I lost the man who made every moment of our 22 years together very special.

Seymour loved life—whether dancing-in his style—or exploring new ideas. Seymour was, indeed, one of a kind.

When Seymour was diagnosed with his illness, he was writing this book. It became clear to him that he would be unable to finish it, so I suggested he ask Wes Johnson to help him. Wes is a former student of Seymour's and they had written a number of papers together. Wes is also a very dear friend. Seymour felt it would be an imposition to ask, but finally, he did. Without hesitation, Wes told Seymour not to worry, that he would finish the book and it would be published.

I knew how important that was to Seymour, for it was one of the goals he would not be able to meet on his own.

For his sensitivity to Seymour's wish, for the technical expertise he brought to the task, and for the years of loving friendship, thank you Wes, from me and Seymour both.

ANNE FLAXMAN GEISSER

SEYMOUR GEISSER

Preface

This book provides a graduate level discussion of four basic modes of statistical inference: (i) frequentist, (ii) likelihood, (iii) Bayesian and (iv) Fisher's fiducial method. Emphasis is given throughout on the foundational underpinnings of these four modes of inference in addition to providing a moderate amount of technical detail in developing and critically analyzing them. The modes are illustrated with numerous examples and counter examples to highlight both positive and potentially negative features. The work is heavily influenced by the work three individuals: George Barnard, Jerome Cornfield and Sir Ronald Fisher, because of the author's appreciation of and admiration for their work in the field. The clear intent of the book is to augment a previously acquired knowledge of mathematical statistics by presenting an overarching overview of what has already been studied, perhaps from a more technical viewpoint, in order to highlight features that might have remained salient without taking a further, more critical, look. Moreover, the author has presented several historical illustrations of the application of various modes and has attempted to give corresponding historical and philosophical perspectives on their development.

The basic prerequisite for the course is a master's level introduction to probability and mathematical statistics. For example, it is assumed that students will have already seen developments of maximum likelihood, unbiased estimation and Neyman-Pearson testing, including proofs of related results. The mathematical level of the course is entirely at the same level, and requires only basic calculus, though developments are sometimes quite sophisticated. There book is suitable for a one quarter, one semester, or two quarter course. The book is based on a two quarter course in statistical inference that was taught by the author at the University of Minnesota for many years. Shorter versions would of course involve selecting particular material to cover.

Chapter 1 presents an example of the application of statistical reasoning by the 12th century theologian, physician and philosopher, Maimonides, followed by a discussion of the basic principles guiding frequentism in Chapter 2. The law of likelihood is then introduced in Chapter 3, followed by an illustration involving the assessment of genetic susceptibility, and then by the various forms of the likelihood principle. Significance testing is introduced and comparisons made between likelihood and frequentist based inferences where they are shown to disagree. Principles of conditionality are introduced.

Chapter 4, entitled "Testing Hypotheses" covers the traditional gamut of material on the Neyman-Pearson (NP) theory of hypothesis testing including most powerful (MP) testing for simple versus simple and uniformly most powerful testing (UMP) for one and two sided hypotheses. A careful proof of the NP fundamental lemma is given. The relationship between likelihood based tests and NP tests is explored through examples and decision theory is introduced and briefly discussed as it relates to testing. An illustration is given to show that, for a particular scenario without the monotone likelihood ratio property, a UMP test exists for a two sided alternative. The chapter ends by showing that a necessary condition for a UMP test to exist in the two sided testing problem is that the derivative of the log likelihood is a non-zero constant.

Chapter 5 discusses unbiased and invariant tests. This proceeds with the usual discussion of similarity and Neyman structure, illustrated with several examples. The sojourn into invariant testing gives illustrations of the potential pitfalls of this approach. Locally best tests are developed followed by the construction of likelihood ratio tests (LRT). An example of a worse-than-useless LRT is given. It is suggested that pre-trial test evaluation may be inappropriate for post-trial evaluation. Criticisms of the NP theory of testing are given and illustrated and the chapter ends with a discussion of the sequential probability ratio test.

Chapter 6 introduces Bayesianism and shows that Bayesian testing for a simple versus simple hypotheses is consistent. Problems with point null and composite alternatives are discussed through illustrations. Issues related to prior selection in binomial problems are discussed followed by a presentation of de Finetti's theorem for binary variates. This is followed by de Finetti's proof of coherence of the Bayesian method in betting on horse races, which is presented as a metaphor for making statistical inferences. The chapter concludes with a discussion of Bayesian model selection.

Chapter 7 gives an in-depth discussion of various theories of estimation. Definitions of consistency, including Fisher's, are introduced and illustrated by example. Lower bounds on the variance of estimators, including those of Cramer-Rao and Bhattacharya, are derived and discussed. The concepts of efficiency and Fisher information are developed and thoroughly discussed followed by the presentation of the Blackwell-Rao result and Bayesian sufficiency. Then a thorough development of the theory of maximum likelihood estimation is presented, and the chapter concludes with a discussion of the implications regarding relationships among the various statistical principles.

The last chapter, Chapter 8, develops set and interval estimation. A quite general method of obtaining a frequentist confidence set is presented and illustrated, followed by discussion of criteria for developing intervals including the concept of conditioning on relevant subsets, which was originally introduced by Fisher. The use of conditioning is illustrated by Fisher's famous "Problem of the Nile." Bayesian interval estimation is then developed and illustrated, followed by development of

Fisher's fiducial inference and a rather thorough comparison between it and Bayesian inference. The chapter and the book conclude with two complex but relevant illustrations, first the Fisher-Behrens problem, which considered inferences for the difference in means for the two sample normal problem with unequal variances, and the second, the Fieller-Creasy problem in the same setting but making inferences about the ratio of two means.

Seymour Geisser received his bachelor's degree in Mathematics from the City College of New York in 1950, and his M.A. and Ph.D. degrees in Mathematical Statistics at the University of North Carolina in 1952 and 1955, respectively. He then held positions at the National Bureau of Standards and the National Institute of Mental Health until 1961. From 1961 until 1965 he was Chief of the Biometry Section at the National Institute of Arthritis and Metabolic Diseases, and also held the position of Professorial Lecturer at the George Washington University from 1960 to 1965. From 1965 to 1970, he was the founding Chair of the Department of Statistics at SUNY, Buffalo, and in 1971 he became the founding Director of the School of Statistics at the University of Minnesota, remaining in that position until 2001. He was a Fellow of the Institute of Mathematical Statistics and the American Statistical Association.

Seymour authored or co-authored 176 scientific articles, discussions, book reviews and books over his career. He pioneered several important areas of statistical endeavor. He and Mervyn Stone simultaneously and independently invented the statistical method called "cross-validation," which is used for validating statistical models. He pioneered the areas of Bayesian Multivariate Analysis and Discrimination, Bayesian diagnostics for statistical prediction and estimation models, Bayesian interim analysis, testing for Hardy-Weinberg Equilibrium using forensic DNA data, and the optimal administration of multiple diagnostic screening tests.

Professor Geisser was primarily noted for his sustained focus on prediction in Statistics. This began with his work on Bayesian classification. Most of his work in this area is summarized in his monograph *Predictive Inference: An Introduction.* The essence of his argument was that Statistics should focus on observable quantities rather than on unobservable parameters that often don't exist and have been incorporated largely for convenience. He argued that the success of a statistical model should be measured by the quality of the predictions made from it.

Seymour was proud of his role in the development of the University of Minnesota School of Statistics and it's graduate program. He was substantially responsible for creating an educational environment that valued the foundations of Statistics beyond mere technical expertise.

Two special conferences were convened to honor the contributions of Seymour to the field of Statistics. The first was held at the National Chiao Tung University of Taiwan in December of 1995, and the second was held at the University of Minnesota in May of 2002. In conjunction with the former conference, a special volume entitled *Modeling and Prediction: Honoring Seymour Geisser*, was published in 1996.

His life's work exemplifies the presentation of thoughtful, principled, reasoned, and coherent statistical methods to be used in the search for scientific truth.

In January of 2004, Ron Christensen and I met with Seymour to tape a conversation with him that has subsequently been submitted to the journal "Statistical Science" for publication. The following quotes are relevant to his approach to the field of statistics in general and are particularly relevant to his writing of "Modes."

- I was particularly influenced by George Barnard. I always read his papers. He had a great way of writing. Excellent prose. And he was essentially trained in Philosophy—in Logic—at Cambridge. Of all of the people who influenced me, I would say that he was probably the most influential. He was the one that was interested in foundations.

- It always seemed to me that prediction was critical to modern science. There are really two parts, especially for Statistics. There is description; that is, you are trying to describe and model some sort of process, which will never be true and essentially you introduce lots of artifacts into that sort of thing. Prediction is the one thing you can really talk about, in a sense, because what you predict will either happen or not happen and you will know exactly where you stand, whether you predicted the phenomenon or not. Of course, Statistics is the so called science of uncertainty, essentially prediction, trying to know something about what is going to happen and what has happened that you don't know about. This is true in science too. Science changes when predictions do not come true.

- Fisher was the master genius in Statistics and his major contributions, in some sense, were the methodologies that needed to be introduced, his thoughts about what inference is, and what the foundations of Statistics were to be. With regard to Neyman, he came out of Mathematics and his ideas were to make Statistics a real mathematical science and attempt to develop precise methods that would hold up under any mathematical set up, especially his confidence intervals and estimation theory. I believe that is what he tried to do. He also originally tried to show that Fisher's fiducial intervals were essentially confidence intervals and later decided that they were quite different. Fisher also said that they were quite different. Essentially, the thing about Neyman is that he introduced, much more widely, the idea of proving things mathematically. In developing mathematical structures into the statistical enterprise.

- Jeffreys had a quite different view of probability and statistics. One of the interesting things about Jeffreys is that he thought his most important contribution was significance testing, which drove [Jerry Cornfield] crazy because, "That's going to be the least important end of statistics." But Jeffreys really brought back the Bayesian point of view. He had a view that you could have an objective type Bayesian situation where you could devise a prior that was more or less reasonable for the problem and, certainly with a large number of observations, the prior would be washed out anyway. I think that was his most important contribution — the rejuvenation of the Bayesian approach before anyone else in statistics through his book, *Probability Theory*. Savage was the one that brought Bayesianism to the States and that is where it spread from.

- My two favorite books, that I look at quite frequently, are Fisher's *Statistical Method in Scientific Inference* and Cramér [*Mathematical Methods of Statistics*]. Those are the two books that I've learned the most from. The one, Cramér, for the mathematics of Statistics, and from Fisher, thinking about the philosophical underpinnings of what Statistics was all about. I still read those books. There always seems to be something in there I missed the first time, the second time, the third time.

In conclusion, I would like to say that it was truly an honor to have been mentored by Seymour. He was a large inspiration to me in no small part due to his focus on foundations which has served me well in my career. He was one of the giants in Statistics. He was also a great friend and I miss him, and his wit, very much. In keeping with what I am quite certain would be his wishes, I would like to dedicate his book for him to another great friend and certainly the one true love of his life, his companion and his occasional foil, his wife Anne Geisser.

The Department of Statistics at the University of Minnesota has established the **Seymour Geisser Lectureship in Statistics**. Each year, starting in the fall of 2005, an individual will be named the Seymour Geisser Lecturer for that year and will be invited to give a special lecture. Individuals will be selected on the basis of excellence in statistical endeavor and their corresponding contributions to science, both Statistical and otherwise. For more information, visit the University of Minnesota Department of Statistics web page, www.stat.umn.edu and click on the SGLS icon.

Finally, Seymour would have wished to thank Dana Tinsley, who is responsible for typing the manuscript, and Barb Bennie, Ron Neath and Laura Pontiggia, who commented on various versions of the manuscript. I thank Adam Branseum for converting Seymour's hand drawn figure to a computer drawn version.

WESLEY O. JOHNSON

CHAPTER ONE

A Forerunner

1.1 PROBABILISTIC INFERENCE—AN EARLY EXAMPLE

An early use of inferred probabilistic reasoning is described by Rabinovitch
(1970).

In the Book of Numbers, Chapter 18, verse 5, there is a biblical injunction which
enjoins the father to redeem his wife's first-born male child by payment of five
pieces of silver to a priest (Laws of First Fruits). In the 12th Century the theologian,
physician and philosopher, Maimonides addressed himself to the following problem with a solution. Suppose one or more women have given birth to a number
of children and the order of birth is unknown, nor is it known how many children
each mother bore, nor which child belongs to which mother. What is the
probability that a particular woman bore boys and girls in a specified sequence?
(All permutations are assumed equally likely and the chances of male or female
births is equal.)

Maimonides ruled as follows: Two wives of different husbands, one primiparous
(P) (a woman who has given birth to her first child) and one not $(\bar{P})$. Let H be the
event that the husband of P pays the priest. If they gave birth to two males (and
they were mixed up), $P(H) = 1$ – if they bore a male (M) and a female
$(F) - P(H) = 0$ (since the probability is only $1/2$ that the primipara gave birth to
the boy). Now if they bore 2 males and a female, $P(H) = 1$.

Case 1	(P)	$(\bar{P})$	Case 2	(P)	$(\bar{P})$
M, M	M	M	M, F	M	F
				F	M
	$P(H) = 1$			$P(H) = \frac{1}{2}$	

Modes of Parametric Statistical Inference, by Seymour Geisser
Copyright © 2006 John Wiley & Sons, Inc.

			PAYMENT	
Case 3	(P)	$(\bar{P})$	*Yes*	*No*
M, M, F	M, M	F	✓	
	F, M	M		✓
	M, F	M	✓	
	F	M, M		✓
	M	F, M	✓	
	M	M, F	✓	

$$P(H) = \tfrac{2}{3}$$

Maimonides ruled that the husband of P pays in Case 3. This indicates that a probability of 2/3 is sufficient for the priest to receive his 5 pieces of silver but 1/2 is not. This leaves a gap in which the minimum probability is determined for payment.

What has been illustrated here is that the conception of equally likely events, independence of events, and the use of probability in making decisions were not unknown during the 12th century, although it took many additional centuries to understand the use of sampling in determining probabilities.

REFERENCES

Rabinovitch, N. L. (1970). Studies in the history of probability and statistics, XXIV Combinations and probability in rabbinic literature. *Biometrika*, **57**, 203–205.

CHAPTER TWO

Frequentist Analysis

This chapter discusses and illustrates the fundamental principles of frequentist-based inferences. Frequentist analysisis and, in particular, significance testing, are illustrated with historical examples.

2.1 TESTING USING RELATIVE FREQUENCY

One of the earliest uses of relative frequency to test a Hypothesis was made by Arbuthnot (1710), who questioned whether the births were equally likely to be male or female. He had available the births from London for 82 years. In every year male births exceeded females. Then he tested the hypothesis that there is an even chance whether a birth is male or female or the probability $p = \frac{1}{2}$. Given this hypothesis he calculated the chance of getting all 82 years of male exceedances $(\frac{1}{2})^{82}$. Being that this is basically infinitesimal, the hypothesis was rejected. It is not clear how he would have argued if some other result had occurred since any particular result is small—the largest for equal numbers of male and female exceedances is less than $\frac{1}{10}$.

2.2 PRINCIPLES GUIDING FREQUENTISM

Classical statistical inference is based on relative frequency considerations. A particular formal expression is given by Cox and Hinkley (1974) as follows:

Repeated Sampling Principle. Statistical procedures are to be assessed by their behavior in hypothetical repetition under the same conditions.

Two facets:

1. Measures of uncertainty are to be interpreted as hypothetical frequencies in long run repetitions.

Modes of Parametric Statistical Inference, by Seymour Geisser
Copyright © 2006 John Wiley & Sons, Inc.

2. Criteria of optimality are to be formulated in terms of sensitive behavior in hypothetical repetitions.

(*Question*: What is the appropriate space which generates these hypothetical repetitions? Is it the sample space S or some other reference set?)

Restricted (Weak) Repeated Sampling Principle. Do not use a procedure which for some possible parameter values gives, in hypothetical repetitions, misleading conclusions most of the time (too vague and imprecise to be constructive). The argument for repeated sampling ensures a physical meaning for the quantities we calculate and that it ensures a close relation between the analysis we make and the underlying model which is regarded as representing the "true" state of affairs.

An early form of frequentist inferences were Tests of Significance. They were long in use before their logical grounds were given by Fisher (1956b) and further elaborated by Barnard (unpublished lectures).

Prior assumption: There is a null hypothesis with no discernible alternatives.

Features of a significance test (Fisher–Barnard)

1. A significance test procedure requires a reference set R (not necessarily the entire sample space) of possible results comparable with the observed result $X = x_0$ which also belongs to R.

2. A ranking of all possible results in R in order of their significance or meaning or departure from a null hypothesis H_0. More specifically we adopt a criterion $T(X)$ such that if $x_1 \succ x_2$ (where x_1 departs further in rank than x_2 both elements of the reference set R) then $T(x_1) > T(x_2)$ [if there is doubt about the ranking then there will be corresponding doubt about how the results of the significance test should be interpreted].

3. H_0 specifies a probability distribution for $T(X)$. We then evaluate the observed result x_0 and the null hypothesis.

$P(T(X) \geq T(x_0) \mid H_0) =$ level of significance or P-value and when this level is small this leads "logically" to a simple disjunction that either:

a) H_0 is true but an event whose probability is small has occurred, or

b) H_0 is false.

Interpretation of the Test:

The test of significance indicates whether H_0 is consistent with the data and the fact that an hypothesis is not significant merely implies that the data do not supply evidence against H_0 and that a rejected hypothesis is very provisional. New evidence is always admissible. The test makes no statement about how the probability of H_0 is made. "No single test of significance by itself can ever establish the existence of H_0 or on the other hand prove that it is false because an event of small probability will occur with no more and no less than its proper frequency, however much we may be surprised it happened to us."

2.3   FURTHER REMARKS ON TESTS OF SIGNIFICANCE

The claim for significance tests are for those cases where alternative hypotheses are not sensible. Note that Goodness-of-Fit tests fall into this category, that is, Do the data fit a normal distribution? Here H_0 is merely a family of distributions rather than a specification of parameter values. Note also that a test of significance considers not only the event that occurred but essentially puts equal weight on more discrepent events that did not occur as opposed to a test which only considers what did occur.

A poignant criticism of Fisherian significance testing is made by Jeffreys (1961). He said

> What the use of the P implies, therefore, is that a hypothesis that may be true may be rejected because it has not predicted observable results that have not occurred.

Fisher (1956b) gave as an example of a pure test of significance the following by commenting on the work of astronomer J. Michell. Michell supposed that there were in all 1500 stars of the required magnitude and sought to calculate the probability, on the hypothesis that they are individually distributed at random, that any one of them should have five neighbors within a distance of a minutes of arc from it. Fisher found the details of Michell's calculation obscure, and suggested the following argument.

> "The fraction of the celestial sphere within a circle of radius a minutes is, to a satisfactory approximation,

$$p = \left(\frac{a}{6875.5}\right)^2,$$

> in which the denominator of the fraction within brackets is the number of minutes in two radians. So, if a is 49, the number of minutes from Maia to its fifth nearest neighbor, Atlas, we have

$$p = \frac{1}{(140.316)^2} = \frac{1}{19689}.$$

> Out of 1499 stars other than Maia of the requisite magnitude the expected number within this distance is therefore

$$m = \frac{1499}{19689} = \frac{1}{13.1345} = .07613.$$

> The frequency with which five stars should fall within the prescribed area is then given approximately by the term of the Poisson series

$$e^{-m}\frac{m^5}{5!},$$

> or, about 1 in 50,000,000, the probabilities of having 6 or more close neighbors adding very little to this frequency. Since 1500 stars each have this probability of being the center of such a close cluster of 6, although these probabilities are not strictly independent,

the probability that among them any one fulfills the condition cannot be far from 30 in a million, or 1 in 33,000. Michell arrived at a chance of only 1 in 500,000, but the higher probability obtained by the calculations indicated above is amply low enough to exclude at a high level of significance any theory involving a random distribution."

With regard to the usual significance test using the "student" t, H_0 is that the distribution is normally distributed with an hypothesized mean $\mu = \mu_0$, and unknown variance σ^2. Rejection can imply that $\mu \neq \mu_0$ or the distribution is not normal or both.

REFERENCES

Arbuthnot, J. (1710). Argument for divine providence taken from the constant regularity of the births of both sexes. *Philosophical Transactions of the Royal Society*, **XXIII**, 186–190.

Cox, D. R. and Hinkley, D. V. (1974). *Theoretical Statistics*. Chapman and Hall, London.

Fisher, R. A. (1956b). *Statistical Methods and Scientific Inference*. Oliver and Boyd.

Jeffreys, H. (1961). *Theory of Probability*. Clarendon Press.

CHAPTER THREE

Likelihood

In this chapter the law of likelihood and other likelihood principles are evoked and issues related to significance testing under different sampling models are discussed. It is illustrated how different models that generate the same likelihood function can result in different frequentist statistical inferences. A simple versus simple likelihood test is illustrated with genetic data. Other principles are also briefly raised and their relationship to the likelihood principle described.

3.1 LAW OF LIKELIHOOD

Another form of parametric inference uses the likelihood—the probability of data D given an hypothesis H or $f(D|H) = L(H|D)$ where H may be varied for given D. A critical distinction of how one views the two sides of the above equation is that probability is a set function while likelihood is a point function.

Law of Likelihood (LL): cf. Hacking (1965) If $f(D|H_1) > f(D|H_2)$ then H_1 is better supported by the data D than is H_2. Hence, when dealing with a probability function indexed by θ, $f(D|\theta) = L(\theta)$ is a measure of relative support for varying θ given D.

Properties of L as a Measure of Support

1. *Transitivity*: Let $H_1 \succ H_2$ indicate that H_1 is better supported than H_2. Then $H_1 \succ H_2$ and $H_2 \succ H_3 \Rightarrow H_1 \succ H_3$.
2. *Combinability*: Relative support for H_1 versus H_2 from independent experiments $\mathcal{E}_1$ and $\mathcal{E}_2$ can be combined, eg. let $D_1 \in \mathcal{E}_1, D_2 \in \mathcal{E}_2, D = (D_1, D_2)$. Then

$$\frac{L_{\mathcal{E}_1, \mathcal{E}_2}(H_1|D)}{L_{\mathcal{E}_1, \mathcal{E}_2}(H_2|D)} = \frac{L_{\mathcal{E}_1}(H_1|D_1)}{L_{\mathcal{E}_1}(H_2|D_1)} \times \frac{L_{\mathcal{E}_2}(H_1|D_2)}{L_{\mathcal{E}_2}(H_2|D_2)}.$$

Modes of Parametric Statistical Inference, by Seymour Geisser
Copyright © 2006 John Wiley & Sons, Inc.

3. *Invariance of relative support under 1–1 transformations g(D):*
Let $D' = g(D)$. For $g(D)$ differentiable and f a continuous probability density

$$L(H|D') = f_{D'}(D'|H) = f_D(D|H)\left|\frac{dg(D)}{dD}\right|^{-1},$$

so

$$\frac{L(H_1|D')}{L(H_2|D')} = \frac{f_{D'}(D'|H_1)}{f_{D'}(D'|H_2)} = \frac{f_D(D|H_1)}{f_D(D|H_2)}.$$

For f discrete the result is obvious.

4. *Invariance of relative support under 1–1 transformation of H:*
Assume H refers to $\theta \in \Theta$ and let $\eta = h(\theta)$ and $h^{-1}(\eta) = \theta$. Then

$$L(\theta|D) = L(h^{-1}(\eta)|D) \equiv \bar{L}(\eta|D).$$

Moreover, with $\eta_i = h(\theta_i)$, $\theta_i = h^{-1}(\eta_i)$,

$$\frac{L(\theta_1|D)}{L(\theta_2|D)} = \frac{L(h^{-1}(\eta_1)|D)}{L(h^{-1}(\eta_2)|D)} = \frac{\bar{L}(\eta_1|D)}{\bar{L}(\eta_2|D)}.$$

Suppose the likelihood is a function of more than one parameter, say $\theta = (\beta, \gamma)$. Consider $H_1 : \beta = \beta_1$ vs. $H_2 : \beta = \beta_2$ while γ is unspecified. Then if the likelihood factors, that is,

$$L(\beta, \gamma) = L(\beta)L(\gamma),$$

then

$$\frac{L(\beta_1, \gamma)}{L(\beta_2, \gamma)} = \frac{L(\beta_1)}{L(\beta_2)}$$

and there is no difficulty. Now suppose $L(\beta, \gamma)$ does not factor so that what you infer about β_1 versus β_2 depends on γ. Certain approximations however may hold if

$$L(\beta, \gamma) = f_1(\beta) f_2(\gamma) f_3(\beta, \gamma)$$

and $f_3(\beta, \gamma)$ is a slowly varying function of β for all γ. Here

$$\frac{L(\beta_1, \gamma)}{L(\beta_2, \gamma)} = \frac{f_1(\beta_1)}{f_1(\beta_2)} \times \frac{f_3(\beta_1, \gamma)}{f_3(\beta_2, \gamma)}$$

and the last ratio on the right-hand side above is fairly constant for β_1 and β_2 and all plausible γ. Then the law of Likelihood for H_1 versus H_2 holds almost irrespective of γ and serves as an approximate ratio. If the above does not hold and $L(\beta, \gamma)$ can be transformed

$$\beta_1 = h_1(\epsilon, \delta), \quad \gamma = h_2(\epsilon, \delta),$$

resulting in a factorization

$$L(h_1, h_2) = L(\epsilon)L(\delta),$$

then likelihood inference can be made on either ϵ or δ separately if they are relevant. Further if this does not hold but

$$L(\beta, \gamma) = L(h_1, h_2) = f_1(\epsilon) f_2(\delta) f_3(\epsilon, \delta),$$

where f_3 is a slowly-varying function of ϵ for all δ then approximately

$$\frac{L(\epsilon_1, \delta)}{L(\epsilon_2, \delta)} = \frac{f_1(\epsilon_1)}{f_2(\epsilon_2)}.$$

A weighted likelihood may also be used as an approximation, namely,

$$\bar{L}(\beta) = \int L(\beta, \gamma)g(\gamma)d\gamma,$$

where $g(\gamma)$ is some "appropriate" weight function and one uses as an approximate likelihood ratio

$$\frac{\bar{L}(\beta_1)}{\bar{L}(\beta_2)}.$$

Other proposals include the profile likelihood,

$$\sup_{\gamma} L(\beta, \gamma) = L(\beta, \hat{\gamma}(\beta, D)),$$

that is, the likelihood is maximized for γ as a function of β and data D. We then compare

$$\frac{L(\beta_1, \hat{\gamma}(\beta_1, D))}{L(\beta_2, \hat{\gamma}(\beta_2, D))}.$$

For further approximations that involve marginal and conditional likelihood see Kalbfleisch and Sprott (1970).

Example 3.1

The following is an analysis of an experiment to test whether individuals with at least one non-secretor allele made them susceptible to rheumatic fever, (Dublin et al., 1964). At the time of this experiment discrimination between homozygous and heterozygous secretors was not possible. They studied offspring of rheumatic secretors (RS) and normal non-secretors (Ns). Table 3.1 presents data discussed in that study.

The simple null and alternative hypotheses considered were:

H_0: Distribution of rheumatic secretors S whose genotypes are Ss or SS by random mating given by the Hardy-Weinberg Law.

H_1: Non-secreting s gene possessed in single or double dose, that is, Ss or ss, makes one susceptible to rheumatic fever, that is, SS not susceptible.

Probabilities for all possible categories calculated under these hypotheses are listed in Table 3.1.

To assess the evidence supplied by the data as to the weight of support of H_0 versus H_1 we calculate:

$$\frac{L(H_0|D)}{L(H_1|D)} = \prod_{\substack{k=1\\k\neq 7,8}}^{9} \frac{p_{k0}^{r_k}(1-p_{k0})^{N_k-r_k}}{p_{k1}^{r_k}(1-p_{k1})^{N_k-r_k}} > 10^9,$$

where

p_{k0} = probability that, out of k offspring from an $RS \times Ns$ family, at least one offspring will be a non-secretor ss given a random mating (Hardy-Weinberg law).

Table 3.1: Secretor Status and Rheumatic Fever

$RS \times Ns$				Expected Proportion	
# of Offspring per Family	# of Families with k Offspring	# of Segregating Families for s	Obs Prob	Random Mating	Susceptible to Rheumatic Fever
k	N_k	r_k	r_k/N_k	$H_0 : p_{k0}$	$H_1 : p_{k1}$
1	16	4	.250	.354	.500
2	32	15	.469	.530	.750
3	21	11	.524	.618	.875
4	11	9	.818	.663	.938
5	5	3	.600	.685	.969
6	3	2	.667	.696	.984
9	1	1	1.000	.706	.998

p_{k1} = probability that, out of k offspring from an $RS \times Ns$ family, at least one offspring will be a non-secretor ss given that all S phenotypes were of the Ss genotype.

From population data, it is known that, among rheumatics, the fraction of secretors S is $P(S) = P(SS) + P(Ss) = 0.701$ and the fraction of nonsecretors is $P(ss) = 0.299$. By applying the Hardy-Weinberg (H-W) law, probabilities for genotypes under random mating are given as:

$$\frac{SS \quad Ss \quad ss}{p^2 \quad 2pq \quad q^2},$$

and thus since $q^2 = 0.299$, we have $q = .54681$ and $p = .45319$, so that $P(S) = 0.20539 + 0.49561 = 0.701$ and moreover, $P(Ss|S) = \frac{P(Ss)}{P(S)} = \frac{0.49561}{0.701} = 0.707$, and similarly $P(SS|S) = 0.293$.

Then under H_0

Relative frequency	$RS \times Ns$	Offspring
.707	$Ss \times ss$	$\frac{1}{2}Ss$ $\frac{1}{2}ss$
.293	$SS \times ss$	Ss

that is, among those families with a secretor (S) and a nonsecretor (ss), there is a 0.707 chance that they will be paired with an Ss and 0.293 probability that they will be paired with an ss. Then the offspring in the former case is equally likely to inherit Ss or ss, and in the latter case, they must inherit Ss. The probability of an ss offspring is thus $0.707 \times \frac{1}{2} = 0.354$. The probability of one or more ss offspring out of k is thus $0.707(1 - (\frac{1}{2})^k)$.

Under H_1,

Under H_1	$RS \times Ns$
$Ss \times ss$	$\frac{1}{2}Ss$ $\frac{1}{2}ss$

So the probability of an ss offspring is $\frac{1}{2}$ and the probability of one or more ss offspring out of k is $1 - (\frac{1}{2})^k$.

3.2   FORMS OF THE LIKELIHOOD PRINCIPLE (LP)

The model for experiment $\mathcal{E}$ consists of a sample space S and a parameter space Θ, a measure μ, and a family of probability functions $f : S \times \Theta \to R^+$ such that for all $\theta \in \Theta$

$$\int_S f d\mu = 1.$$

1. *Unrestricted LP* (ULP)

   For two such experiments modeled as $\mathcal{E} = (S, \mu, \Theta, f)$ and $\mathcal{E}' = (S', \mu', \Theta, f')$, and for realization $D \in S$ and $D' \in S'$, if

   $$f(D|\theta) = g(D, D')\, f'(D'|\theta) \quad \text{for } g > 0$$

   for all θ and the choice of $\mathcal{E}$ or $\mathcal{E}'$ is uniformative with regard to θ, then the evidence or inferential content or information concerning θ, all denoted by Inf is such that

   $$\text{Inf}(\mathcal{E}, D) = \text{Inf}(\mathcal{E}', D').$$

   Note this implies that all of the statistical evidence provided by the data is conveyed by the likelihood function. There is an often useful extension, namely, when $\theta = (\theta_1, \theta_2)$, $\delta_1 = h(\theta_1, \theta_2)$, $\delta_2 = k(\theta_1, \theta_2)$, and

   $$f(D|\theta) = g(D, D', \delta_1)\, f'(D'|\delta_2),$$

   then $\text{Inf}(\mathcal{E}, D) = \text{Inf}(\mathcal{E}', D')$ concerning δ_2.

2. *Weakly restricted LP* (RLP)

   LP is applicable whenever a) $(S, \mu, \Theta, f) = (S', \mu', \Theta, f')$ and b) $(S, \mu, \Theta, f) \neq (S', \mu', \Theta, f')$ when there are no structural features of (S, μ, Θ, f) which have inferential relevance and which are not present in (S', μ', Θ, f').

3. *Strongly restricted LP* (SLP)

   Applicable only when $(S, \mu, \Theta, f) = (S', \mu', \Theta, f')$.

4. *Extended LP* (ELP)

   When $\theta = (p, \gamma)$ and $f(D, D'|p, \gamma) = g(D, D'|p)f'(D'|\gamma)$ it is plausible to extend LP to

   $$\text{Inf}(\mathcal{E}, D) = \text{Inf}(\mathcal{E}', D'),$$

   concerning p assuming p and γ are unrelated.

To quote Groucho Marx, "These are my principles and if you don't like them I have others," see Sections 3.6 and 7.12.

In summary LP and law of likelihood (LL) assert that all the information or evidence which data provide concerning the relative plausibility of H_1 and H_2 is contained in the likelihood function and the ratio is to be defined as the degree to which H_1 is supported (or the plausibility) relative to H_2 given D with the caution concerning which form of LP is applicable.

The exposition here leans on the work of Barnard et al. (1962), Barnard and Sprott (1971), Hacking (1965), and Birnbaum (1962).

3.3 LIKELIHOOD AND SIGNIFICANCE TESTING

We now compare the use of likelihood analysis with a significance test. Suppose we are only told that in a series of independent and identically distributed binary trials there were r successes and $n - r$ failures, and the sampling was conducted in one of three ways:

1. The number of trials was fixed at n.
2. Sampling was stopped at the rth success.
3. Sampling was stopped when $n - r$ failures were obtained.

Now while the three sampling probabilities differ they all have the same likelihood:

$$L = p^r(1 - p)^{n-r}.$$

The probabilities of r successes and $n - r$ failures under these sampling methods are:

$$f_n = \binom{n}{r} L \quad r = 0, 1, 2, \ldots, n$$

$$f_r = \binom{n-1}{r-1} L \quad n = r, r+1, \ldots$$

$$f_{n-r} = \binom{n-1}{n-r-1} L \quad n = r+1, r+2, \ldots$$

where f_a denotes the probability where subscript a is fixed for sampling.

Example 3.2

Suppose the data given are $r = 1$, $n = 5$. Then a simple test of significance for $H_0 : p = \frac{1}{2}$ vs. $p < \frac{1}{2}$ depends critically on the sampling distribution, since

$$P_n = \binom{5}{0}\left(\frac{1}{2}\right)^5 + \binom{5}{1}\left(\frac{1}{2}\right)^5 = \frac{6}{32} = \frac{3}{16};$$

$$P_r = P[N \geq 5 | r = 1] = 1 - P(N \leq 4 | r = 1) = 1 - \sum_{n=1}^{4} P(n | r = 1)$$

$$= 1 - [P[N = 1] + P[N = 2] + P[N = 3] + P(N = 4)]$$

$$= 1 - \left[\frac{1}{2} + \binom{1}{0}\left(\frac{1}{2}\right)^2 + \binom{2}{0}\left(\frac{1}{2}\right)^3 + \binom{3}{0}\left(\frac{1}{2}\right)^4\right]$$

$$= 1 - \frac{1}{2} - \frac{1}{4} - \frac{1}{8} - \frac{1}{16} = \frac{1}{16};$$

$$P_{n-r} = \binom{n-1}{n-r-1} L = \binom{n-1}{r} L, \quad n \geq r+1$$

$$= P[N \geq 5] = 1 - P[N \leq 4] = 1 - \sum_{n=2}^{4} P(n|r=1)$$

$$= 1 - \left[\binom{1}{0}\left(\frac{1}{2}\right)^2 + \binom{2}{1}\left(\frac{1}{2}\right)^3 + \binom{3}{2}\left(\frac{1}{2}\right)^4 \right]$$

$$= 1 - \left(\frac{1}{4} + \frac{2}{8} + \frac{3}{16}\right) = 1 - \frac{11}{16} = \frac{5}{16}.$$

Here significance testing provides three different P values but adherence to LP would require making the same inference assuming no structural features having inferential relevance.

3.4 THE 2 × 2 TABLE

If we are dealing with the classical 2×2 table, then the random values within Table 3.2 have the multinomial probability function

$$f_n = f(r_1, r_2, n_1 - r_1, n_2 - r_2)$$

$$= \binom{n}{r_1, r_2, n_1 - r_1, n_2 - r_2} p_{11}^{r_1} p_{12}^{n_1 - r_1} p_{21}^{r_2} p_{22}^{n_2 - r_2}, \tag{3.4.1}$$

subject to the four arguments summing to n and $\sum_{i,j} p_{ij} = 1$, with prescribed sample size n. Let

$$p_1 = \frac{p_{11}}{p_{11} + p_{12}}, \quad p_2 = \frac{p_{21}}{p_{21} + p_{22}}.$$

Table 3.2: Classical 2 × 2 Table

		A		$\bar{A}$		
B		p_{11}		p_{12}		$p_1.$
	r_1		$n_1 - r_1$		n_1	
$\bar{B}$		p_{21}		p_{22}		$p_2.$
	r_2		$n_2 - r_2$		n_2	
	r	$p_{.1}$	$n - r$	$p_{.2}$	n	1

I. *Independence:* $p_{11} = p_1.p._1$, $p_{12} = p_1.p._2$, $p_{21} = p_2.p._1$, $p_{22} = p_2.p._2$
II. *Equality:* $p_1 = p_2$

It is easy to show that *I* holds if and only if *II* holds.

Large Sample Test—Chi-Square test

$$X^2 = n[r_1(n_2 - r_2) - r_2(n_1 - r_1)]^2 / r n_1 n_2 (n - r)$$

is χ^2 with one degree of freedom asymptotically whether the sampling distribution is multinomial or independent binomials. Note that

$$
f_n = f(r_1, r_2 | n_1, n_2) f(n_1 | n)
$$
$$
= \binom{n_1}{r_1} p_1^{r_1} (1 - p_1)^{n_1 - r_1} \binom{n_2}{r_2} p_2^{r_2} (1 - p_2)^{n_2 - r_2} \binom{n}{n_1} \gamma^{n_1} (1 - \gamma)^{n_2}, \tag{3.4.2}
$$

where $\gamma = p_{11} + p_{12} = p_1$.

Inference about a function, $g(p_1, p_2)$, will be the same from f_n as from the first two factors of the right-hand side of (3.4.2) by invoking the extension of the usual Likelihood Principle, ELP.

If we now condition one of the other marginal sums, say $r = r_1 + r_2$, then $n - r$ is also fixed and we have a conditional on all of the marginals. This yields

$$
f(r_1 | r, n_1, n_2) = \binom{n_1}{r_1} \binom{n_2}{r - r_1} \Psi^{r_1} \bigg/ \sum_{j=a}^{b} \binom{n_1}{j} \binom{n_2}{r - j} \Psi^{j}, \tag{3.4.3}
$$

where $b = \min(r, n_1)$ and $a = \max(r - n_2, 0)$ and $\Psi = p_1(1 - p_2)/p_2(1 - p_1)$. Note that $\Psi = 1$ iff $p_1 = p_2$.

This was originally proposed by Fisher who provided us with the exact test under $H_0 : \Psi = 1$ determined from the hypergeometric probability function,

$$
f(r_1) = \binom{n_1}{r_1} \binom{n_2}{r - r_1} \bigg/ \binom{n}{r}. \tag{3.4.4}
$$

Note that the likelihood for the independent binomials depends on two parameters and cannot be factored, that is, for *IB* representing independent binomials,

$$
f_{IB} = \binom{n_1}{r_1} \binom{n_2}{r_2} \Psi^{r_1} (1 - p_2)^{n-r} p_2^r \bigg/ \left(1 + \frac{p_2 \Psi}{1 - p_2} \right)^{n_1} \tag{3.4.5}
$$

and therefore LP is violated when using (3.4.3). Hence Fisher's exact test loses some information if the sampling situation started out as a multinomial or two independent binomials.

Table 3.3: A 2 × 2 Table

	S	D	
T	3	0	3
C	0	3	3
	3	3	

Barnard (1945) provided the example shown in Table 3.3, where $S =$ survived, $D =$ died, $T =$ treatment, $C =$ control, and where $H_0 : T = C$ versus $H_1 : T > C$. For this table, assuming sampling was from two independent binomial distributions where $H_0 : p_1 = p_2$ we obtain

$$f(r_1, r_2|p) = \binom{3}{r_1}\binom{3}{r_2}p^r(1-p)^{6-r}.$$

The possible events are given in Table 3.4 for this type of sample: Events 4, 7, 10, and 13 belong to the reference set associated with the hypergeometric probability function, and

$$P(13) = \frac{1}{\binom{6}{3}} = \frac{1}{20}. \tag{3.4.6}$$

Now for independent binomials,

$$P(13) = p^3(1-p)^3 \le \left(\frac{1}{2}\right)^6 = \frac{1}{64}. \tag{3.4.7}$$

So one might argue that the P-value was not more than $\frac{1}{64}$. Although Barnard (1945) originally proposed this view he later recanted in favor of the hypergeometric P-value of $\frac{1}{20}$, which was favored by Fisher.

Consider the following problem: A laboratory claims its diagnoses are better than random. The laboratory is tested by presenting it with n test specimens and is told that n_1 of the n, are positive and n_2 are negative. The laboratory will divide its results such that n_1 are positive and n_2 are negative. Suppose their results say that r_1 of the n_1 are positive and $n_1 - r_1$ are negative, while among the negative n_2 the laboratory will divide its results as saying $n_1 - r_1$ are positive which leaves $n_2 - n_1 + r_1$ as negative. Table 3.5 illustrates the situation. The null and alternative hypotheses are: H_0: the results are random; H_1: the results are better than random.

Under H_0:

$$P(r_1) = \frac{\binom{n_1}{r_1}\binom{n_2}{n_1 - r_1}}{\binom{n}{n_1}}.$$

Table 3.4: 2 × 2 Tables

	(1)		(2)		(3)		(4)			
	S	D	S	D	S	D	S	D		
T	0	3	0	3	0	3	0	3	3	
C	0	3	1	2	2	1	3	0	3	
	0	6	1	5	2	4	3	3	6	
# of ways	1		3		3		1			8
Probability	$(1-p)^6$		$p(1-p)^5$		$p^2(1-p)^4$		$p^3(1-p)^3$			
	(5)		(6)		(7)		(8)			
T	1	2	1	2	1	2	1	2	3	
C	0	3	1	2	2	1	3	0	3	
	1	5	2	4	3	3	4	2	6	
# of ways	3		9		9		3			24
Probability	$p(1-p)^5$		$p^2(1-p)^4$		$p^3(1-p)^3$		$p^4(1-p)^2$			
	(9)		(10)		(11)		(12)			
T	2	1	2	1	2	1	2	1	3	
C	0	3	1	2	2	1	3	0	3	
	2	4	3	3	4	2	5	1	6	
# of ways	3		9		9		3			24
Probability	$p^2(1-p)^4$		$p^3(1-p)^3$		$p^4(1-p)^2$		$p^5(1-p)$			
	(13)		(14)		(15)		(16)			
T	3	0	3	0	3	0	3	0	3	
C	0	3	1	2	2	1	3	0	3	
	3	3	4	2	5	1	6	0	6	
# of ways	1		3		3		1			8
Probability	$p^3(1-p)^3$		$p^4(1-p)^2$		$p^5(1-p)$		p^6			

For $n_1 = 5$, $n_2 = 3$, $r_1 = 4$, $r_2 = 1$, we obtain

$$P(r_1 = 5) = \frac{\binom{5}{5}\binom{3}{0}}{\binom{8}{5}} = \frac{1}{56}, \quad P(r_1 = 4) = \frac{\binom{5}{4}\binom{3}{1}}{\binom{8}{5}} = \frac{15}{56},$$

$$P(r_1 = 3) = \frac{\binom{5}{3}\binom{3}{2}}{\binom{8}{5}} = \frac{30}{56}, \quad P(r_1 = 2) = \frac{\binom{5}{2}\binom{3}{3}}{\binom{8}{5}} = \frac{10}{56}.$$

Table 3.5: Laboratory Positives and Negatives

		Laboratory reports		
		positive	negative	
	positive	r_1	$n_1 - r_1$	n_1
True				
	negative	$n_1 - r_1$	$n_2 - n_1 + r_1$	n_2
		n_1	n_2	n

Observing $r_1 = 5$ yields $P < 0.02$ provides evidence for rejecting H_0.
Fisher's Tea–Taster (1935): A woman claimed she could tell whether the milk (M) was poured into the tea (T) or the tea into the milk. Fisher's proposed test was to have 4 cups such that T $\longrightarrow$ M and 4 cups M $\longrightarrow$ T. The Tea Taster was told the situation and she was to divide the cups. Table 3.6 illustrates this.
   We calculate the various possibilities,

$$P(4) = \frac{\binom{4}{4}\binom{4}{0}}{\binom{8}{4}} = \frac{1}{70}, \quad P(3) = \frac{\binom{4}{3}\binom{4}{1}}{70} = \frac{16}{70}, \quad P(2) = \frac{\binom{4}{2}\binom{4}{2}}{70} = \frac{36}{70},$$

$$P(1) = \frac{\binom{4}{1}\binom{4}{3}}{70} = \frac{16}{70}, \quad P(0) = \frac{\binom{4}{0}\binom{4}{4}}{70} = \frac{1}{70}.$$

If the Tea–Taster correctly identified all the cups, the chance of this happening under a null hypothesis of random guessing is $P = \frac{1}{70}$ which presumably would provide evidence against the null hypothesis of random guessing.

3.5 SAMPLING ISSUES

There are many other ways of sampling that can lead to a 2 × 2 table. For example, we can allow n to be random (negative multinomial sampling) and condition on any

Table 3.6: The Lady Tasting Tea

		Tea–Taster Assertion		
		T $\rightarrow$ M	M $\rightarrow$ T	
Actual	T $\rightarrow$ M	r_1	$4 - r_1$	4
Occurrence	M $\rightarrow$ T	$4 - r_1$	r_1	4
		4	4	8

one of the marginals or tabular entries. Suppose then for n random we sample until a fixed value of n_1 is achieved. We then find

$$f_{n1} = f(r_1, r_2, n \mid n_1) = f(r_1, r_2 \mid n_1, n) f(n \mid n_1)$$

$$= \binom{n_1}{r_1} p_1^{r_1} (1 - p_1)^{n_1 - r_1} \binom{n_2}{r_2} p_2^{r_2} (1 - p_2)^{n_2 - r_2} \binom{n-1}{n_1 - 1} \gamma^{n_1} (1 - \gamma)^{n_2}.$$

$$(3.5.1)$$

However, the likelihood for p_1 and p_2 is still the same, although the overall sampling distribution is obviously different than the usual multinomial. Hence inference about functions of p_1 and p_2, according to the ELP, is still the same as when we assumed multinomial sampling.

Negative multinomial sampling can also occur if one sampled n until a fixed r is achieved. In this case we get

$$f_r = f(r_1, n_1, n_2 \mid r) = f(n_1, n_2 \mid r_1, r_2) f(r_1 \mid r)$$

$$= \binom{n_1 - 1}{r_1 - 1} p_1^{r_1} (1 - p_1)^{n_1 - r_1} \binom{n_2 - 1}{r_2 - 1} p_2^{r_2} (1 - p_2)^{n_2 - r_2} \binom{r}{r_1} \alpha^{r_1} (1 - \alpha)^{r_2},$$

$$(3.5.2)$$

where $\alpha = p_{11}/(p_{11} + p_{21})$.

Although the likelihood for p_1 and p_2 arises from two independent negative binomials it is the same as in the positive multinomial and the independent binomials case. However, a frequentist can condition on $n_1 + n_2$ yielding a sampling probability function

$$f(n_1 \mid r_1, r_2, n) = \binom{n_1 - 1}{r_1 - 1}$$

$$\times \binom{n - n_1 - 1}{r_2 - 1} \theta^{n_1} \Bigg/ \sum_{j=a}^{b} \binom{j-1}{r_1 - 1} \binom{n-j-1}{r_2 - 1} \theta^{j}, \quad (3.5.3)$$

where $a = \min(r_1, n - r_2)$, $b = n - r_2$ and $\theta = (1 - p_1/1 - p_2)$, that is, the ratio of the failure probabilities. Here the parametrization differs from (3.4.3–3.4.5) and the likelihood from (3.5.2) which is also the likelihood for independent negative binomials. Again the ELP is not sustained. Since $\theta = 1$ is equivalent to $p_1 = p_2$, we have a frequentist significance test based on the negative hypergeometric distribution,

$$f(n_1 \mid r_1, r_2, n) = \frac{\binom{n_1 - 1}{r_1 - 1} \binom{n - n_1 - 1}{r_2 - 1}}{\binom{n-1}{r_1 + r_2 - 1}}. \quad (3.5.4)$$

The "Mixed" Sampling Case

Another negative multinomial sampling approach stops when r_1, say, attains a given value. Here

$$
\begin{aligned}
f_{r_1} &= f(r_2, n_1, n \mid r_1) = f(r_2, n \mid n_1, r_1) f(n_1 \mid r_1) \\
&= f(n_1 \mid r_1) f(r_2 \mid n, n_1, r_1) f(n \mid n_1, r_1) \\
&= \binom{n_1 - 1}{r_1 - 1} p_1^{r_1} (1 - p_1)^{n_1 - r_1} \binom{n_1}{r_2} \\
&\quad \times p_2^{r_2} (1 - p_2)^{n_2 - r_2} \binom{n - 1}{n_1 - 1} \gamma^{n_1} (1 - \gamma)^{n - n_1}.
\end{aligned}
\tag{3.5.5}
$$

Again, the likelihood for p_1 and p_2 is preserved for inference that respects ELP but here we now encounter a difficulty for conditional frequentist inference regarding p_1 and p_2. What does one condition on to obtain an exact significance test on $p_1 = p_2$? Of course, this problem would also occur when we start with one sample that is binomial, say, a control, and the other negative binomial, for say a new treatment where one would like to stop the latter trial after a given number of failures. Note that this problem would persist for f_{r_2}, $f_{n_1 - r_1}$ and $f_{n_2 - r_2}$. Hence in these four cases there is no exact conditional Fisher type test for $p_1 = p_2$.

Next we examine these issues for the 2×2 table. Here we list the various ways one can sample in constructing a 2×2 table such that one of the nine values is fixed, that is, when that value appears sampling ceases. For 7 out of the 9 cases the entire likelihood is the same, where

$$
L = \gamma^{n_1} (1 - \gamma)^{n_2} \prod_{i-1}^{2} p_i^{r_i} (1 - p_i)^{n_i - r_i} = L(\gamma) L(p_1, p_2)
\tag{3.5.6}
$$

with sampling probabilities

$$
f_n = \binom{n_1}{r_1} \binom{n_2}{r_2} \binom{n}{n_1} L,
$$

$$
f_{n_1} = \binom{n_1}{r_1} \binom{n_2}{r_2} \binom{n - 1}{n_1 - 1} L = f(r_1, r_2, n \mid n_1)
$$

$$
= f(r_1, r_2 \mid n, n_1) f(n \mid n_1),
$$

$$
f_{n_2} = \binom{n_1}{r_1} \binom{n_2}{r_2} \binom{n - 1}{n_2 - 1} L = f(r_1, r_2 \mid n_2, n) f(n \mid n_2),
$$

$$f_{r_1} = \binom{n_1 - 1}{r_1 - 1}\binom{n_2}{r_2}\binom{n - 1}{n_1 - 1} L = f(r_2, n_1, n|r_1)$$

$$= f(r_2, n|n_1, r_1)f(n_1|r_1)$$

$$= f(r_2|n_1, r_1, n_2)f(n|n_1, r_2) \cdot f(n_1|r_1),$$

$$f_{r_2} = \binom{n_1}{r_1}\binom{n_2 - 1}{r_2 - 1}\binom{n - 1}{n_2 - 1} L,$$

$$f_{n_1 - r_1} = \binom{n_1 - 1}{n_1 - r_1 - 1}\binom{n_2}{r_2}\binom{n - 1}{n_1 - 1} L,$$

$$f_{n_2 - r_2} = \binom{n_1}{r_1}\binom{n_2 - 1}{n_2 - r_2 - 1}\binom{n - 1}{n_2 - 1} L.$$

The other two, whose total likelihoods differ from the above, are still equivalent to the above for inference on (p_1, p_2) by the virtue of the ELP.

$$f_r = \binom{n_1 - 1}{r_1 - 1}\binom{n_2 - 1}{r_2 - 1}\binom{r}{r_1} L(p_1, p_2)\alpha^{r_1}(1 - \alpha)^{r_2} = f(r_1, n_1, n_2|r)$$

$$= f(n_1, n_2|r_1, r_2) f(r_1|r),$$

$$f_{n-r} = \binom{n_1 - 1}{n_1 - r_1 - 1}\binom{n_2 - 1}{n_2 - r_2 - 1}\binom{n - r}{n - r_1}$$

$$\times L(p_1, p_2)\beta^{n_1 - r_1}(1 - \beta)^{n_2 - r_2}.$$

3.6 OTHER PRINCIPLES

1. *Restricted Conditionality Principle RCP*: Same preliminaries as LP.
   $\mathcal{E} = (S, \mu, \theta, f)$ is a mixture of experiments $\mathcal{E}_i = (S_i, \mu_i, \theta, f_i)$ with mixture probabilities q_i independent of θ. First we randomly select $\mathcal{E}_1$ or $\mathcal{E}_2$ with probabilities q_1 and $q_2 = 1 - q_1$, and then perform the chosen experiment $\mathcal{E}_i$. Then we recognize the sample $D = (i, D_i)$ and $f(D|\theta) = q_i f_i(D_i|\theta), i = 1, 2$. Then RCP asserts $\mathrm{Inf}(\mathcal{E}, D) = \mathrm{Inf}(\mathcal{E}_i, D_i)$.

   Definition of Ancillary Statistic: A statistic $C = C(D)$ is ancillary with respect to θ if $f_C(c|\theta)$ is independent of θ, so that an ancillary is non-informative about θ. $C(D)$ maps $S \to S_c$ where each $c \in S_c$ defines $S_c = (D|C(D) = c)$. Define the conditional experiment $\mathcal{E}_{D|C} = (S_c, \mu, \theta, f_{D|C}(D|c))$, and the marginal experiment $\mathcal{E}_C = (S_c, \mu, \theta, f_C(c))$, where $\mathcal{E}_C = $ sample from S_c or sample from S and observe c and $\mathcal{E}_{D|C} = $ conditional on $C = c$, sample from S_c.

2. *Unrestricted Conditionality Principle (UCP)*: When C is an ancillary $\mathrm{Inf}(\mathcal{E}, D) = \mathrm{Inf}(\mathcal{E}_{D|C}, D)$ concerning θ. It is as if we performed $\mathcal{E}_C$ and then performed $\mathcal{E}_{D|C}$.

3. *Mathematical Equivalence Principle (MEP)*: For a single $\mathcal{E}$, if f for all $\theta \in \Theta$ is such that $f(D|\theta) = f(D'|\theta)$ then

$$\text{Inf}(\mathcal{E}, D) = \text{Inf}(\mathcal{E}, D').$$

Note this is just a special case of ULP.

We show that ULP $\Leftrightarrow$ (RCP, MEP). First assume ULP, so that $\text{Inf}(\mathcal{E}, D) = \text{Inf}(\mathcal{E}', D')$. Now suppose $f(D|\theta) = f(D'|\theta)$, then apply ULP so $\text{Inf}(\mathcal{E}, D') = \text{Inf}(\mathcal{E}, D')$, which is MEP. Further suppose, where C is an ancillary, and hence that

$$f(D|\theta) = f(D|c, \theta)h(c), \quad f(D'|\theta) = f(D'|c, \theta)h(c).$$

Hence ULP implies that

$$\text{Inf}(\mathcal{E}, D) = \text{Inf}(\mathcal{E}, D|C)$$

or UCP, and UCP implies RCP.

Conversely, assume (RCP, MEP) and that $(\mathcal{E}_1, D_1)$ and $(\mathcal{E}_2, D_2)$ generate equivalent likelihoods

$$f_1(D_1|\theta) = h(D_1, D_2)f_2(D_2|\theta).$$

We will use (RCP, MEP) to show $\text{Inf}(\mathcal{E}_1, D_1) = \text{Inf}(\mathcal{E}_2, D_2)$. Now let $\mathcal{E}$ be the mixture experiment with probabilities $(1 + h)^{-1}$ and $h(1 + h)^{-1}$. Hence the sample points in $\mathcal{E}$ are $(1, D_1)$ and $(2, D_2)$ and

$$f(1, D_1|\theta) = (1 + h)^{-1}f_1(D_1|\theta) = h(1 + h)^{-1}f_2(D_2|\theta),$$
$$f(2, D_2|\theta) = h(1 + h)^{-1}f_2(D_2|\theta).$$

Therefore $f(1, D_1|\theta) = f(2, D_2|\theta)$ so from MEP

$$\text{Inf}(\mathcal{E}, (1, D_1)) = \text{Inf}(\mathcal{E}, (2, D_2)).$$

Now apply RCP to both sides above so that

$$\text{Inf}(\mathcal{E}_1, D_1) = \text{Inf}(\mathcal{E}, \theta, (1, D_1)) = \text{Inf}(\mathcal{E}, \theta, (2, D_2)) = \text{Inf}(\mathcal{E}_2, D_2),$$
therefore (RCP, MEP) $\Longrightarrow$ ULP

REFERENCES

Barnard, G. A. (1945). A new test for 2×2 tables. *Nature*, **156**, 177.

Barnard, G. A., Jenkins, G. M., and Winsten, C. B. (1962). Likelihood inference and time series. *Journal of the Royal Statistical Society A*, **125**, 321–372.

Barnard, G. A. and Sprott, D. A. (1971). A note on Basu's examples of anomalous ancillary statistics (see reply, p 175). *Foundations of Statistical Inference*, ed. Godambe, V. P. and Sprott, D. A., Holt, Rinehart and Winston, New York, pp. 163–170.

Birnbaum, A. (1962). On the foundations of statistical inference. *Journal of the American Statistical Association*, **57**, 269–326.

Dublin, T. et al. (1964). Red blood cell groups and ABH secretor system as genetic indicators of susceptibility to rheumatic fever and rheumatic heart disease. *British Medical Journal*, September 26, **vol. ii**, 775–779.

Fisher, R. A. (1960). *The Design of Experiments*. Oliver and Boyd, Edinburgh.

Hacking, I. (1965). *Logic of Statistical Inference*. Cambridge University Press.

Kalbfleisch, J. D. and Sprott, D. A. (1970). Application of likelihood methods to models involving large numbers of parameters. *Journal of the Royal Statistical Society*, B, **32**, 175–208.

C H A P T E R   F O U R

Testing Hypotheses

This chapter discusses the foundational aspects of frequentist hypothesis testing. The Neyman-Pearson theory of most powerful (MP) and uniformly most powerful (UMP) tests is developed. Simple illustrations are given as examples of how the theory applies and also to show potential problems associated with the frequentist methodology. The concept of risk function is introduced and applied to the testing scenario. An illustration of a UMP test for a point null with a two sided alternative in a model without the monotone likelihood ratio (MLR) property is given. A necessary condition for the existence of a UMP test for a two-sided alternative with a point null is derived.

4.1 HYPOTHESIS TESTING VIA THE REPEATED SAMPLING PRINCIPLE

Neyman-Pearson (N-P) (1933) Theory of Hypotheses Testing is based on the repeated sampling principle and is basically a two decision procedure.

We will collect data D and assume we have two rival hypotheses about the populations from which the data were generated: H_0 (the null hypothesis) and H_1 (the alternative hypothesis). We assume a sample space S of all possible outcomes of the data D. A rule is then formulated for the rejection of H_0 or H_1 in the following manner. Choose a subset s (critical region) of S and

if $D \in s$ reject H_0 and accept H_1

if $D \in S - s$ reject H_1 and accept H_0

according to

$P(D \in s | H_0) = \epsilon$ (size) $\leq \alpha$ (level) associated with the test (Type 1 error)

$P(D \in s | H_1) = 1 - \beta$ (power of the test)

$P(D \in S - s | H_1) = \beta$ (Type 2 error)

Modes of Parametric Statistical Inference, by Seymour Geisser
Copyright © 2006 John Wiley & Sons, Inc.

The two basic concepts are size and power and N-P theory dictates that we choose a test (critical region) which results in small size and large power. At this juncture we assume size equals level and later show how size and level can be equated. Now

$$\alpha = P(\text{accepting } H_1|H_0) \quad \text{or} \quad 1 - \alpha = P(\text{accepting } H_0|H_0)$$

and

$$1 - \beta = P(\text{accepting } H_1|H_1) \quad \text{or} \quad \beta = P(\text{accepting } H_0|H_1).$$

Since there is no way of jointly minimizing size and maximizing power, N-P suggest choosing small size α and then maximize power $1 - \beta$ (or minimize β).

4.2 REMARKS ON SIZE

Before you have the data D it would be reasonable to require that you should have small size (frequency of rejecting H_0 when H_0 is true). But this may mislead you once you have the data, for example, suppose you want size $= .05$ where the probabilities for the data $D = (D_1, D_2, D_3)$ are given in Table 4.1.

Presumably if you want size $= .05$ you reject H_0 if event D_1 occurs and accept if D_2 or D_3 occur. However if D_1 occurs you are surely wrong to reject H_0 since $P(D_1|H_1) = 0$. So you need more than size. Note that before making the test, all tests of the same size provide us with the same chance of rejecting H_0, but after the data are in hand not all tests of the same size are equally good. In the N-P set up we are forced to choose a test before we know what the sample value actually is, even when our interest is in evaluating hypotheses with regard to the sample data we have. Therefore if two tests T_1 and T_2 have the same size one might be led to choose the test with greater power. That this is not necessarily the best course is demonstrated in the following Example 4.1 and its variations, Hacking (1965).

Example 4.1

Let tests T_1 and T_2 have the same size for the setup of Table 4.2.

Let T_1 reject H_0 if D_3 occurs: size $= 0.01 = P(D_3|H_0)$, power $= 0.97 = P(D_3|H_1)$.

Let T_2 reject H_0 if D_1 or D_2 occur: size $= 0.01 = P(D_1 \text{ or } D_2|H_0)$, power $= 0.02 = P(D_1 \text{ or } D_2|H_1)$.

Table 4.1: Probabilities for Data under H_0 and H_1

$P(D_1|H_0) = .05$	$P(D_1|H_1) = 0$
$P(D_2|H_0) = .85$	$P(D_2|H_1) = .85$
$P(D_3|H_0) = .10$	$P(D_3|H_1) = .15$

Table 4.2: Probabilities under H_0 and H_1

$D =$	D_1	D_2	D_3	D_4
H_0:	0	.01	.01	.98
H_1:	.01	.01	.97	.01

Is T_1 really a better test than T_2 after an event D_1 has occurred? Surely not, because after D_1 occurred it is clear T_2 is the better test although before any D is observed T_1 is far superior.

The test that is better before the data are collected need not be better after the data are in hand. N-P did not stop with size and power alone. N-P theory states that if a most powerful test exists it is the best.

Suppose among all regions (sets) s^* having size ϵ, that is,

$$P(D \in s^*|H_0) = \epsilon,$$

that s is such that

$$P(D \in s^*|H_1) \le P(D \in s|H_1).$$

Hence for a given ϵ, s is that region that maximizes power. Consider all possible tests of size $= 0.01$ in Table 4.3. Therefore T_4 is most powerful among tests of size 0.01 and is best before or after D is observed.

Suppose we have a composite alternative say $K = \{H_1, H_2\}$ where now the setup is as shown in Table 4.4. Consider again all tests of size 0.01 in Table 4.5. Here T_4 is now the uniformly most powerful (UMP) test of size 0.01. However, note what happens if D_2 occurs. Using T_4, you accept H_0 and reject H_2 even though the support for H_2 is 1.9 times greater than for H_0. Should you then reject H_2 if D_2 has occurred?

4.3   UNIFORMLY MOST POWERFUL TESTS

Suppose H and K are both composite hypotheses. Then if for all s^* satisfying

$$P(D \in s^*|H) \le \alpha \text{ (level of the test) for all } H,$$

Table 4.3: Size and Power H_0 versus H_1

Test	Critical Region	Size	Power under H_1
T_1	reject H_0 if D_3	0.01	0.97
T_2	reject H_0 if D_1 or D_2	0.01	0.02
T_3	reject H_0 if D_2	0.01	0.01
T_4	reject H_0 if D_3 or D_1	0.01	0.98

Table 4.4: Probabilities under H_0, H_1, and H_2

	D_1	D_2	D_3	D_4
H_0	0	.01	.01	.98
H_1	.01	.01	.97	.01
H_2	.001	.019	.97	.01

and if s is such that

$$P(D \in s^*|K) \leq P(D \in s|K) \quad \text{for all } K,$$

then the test based on s is UMP. Often H is a simple H_0 and $K = H_1$ depends on a parameter that is defined over an interval. Consider the following example, for $0 < p < 1$:

$$P(R = r|p) = \binom{n}{r} p^r (1 - p)^{n-r}.$$

Denote the likelihood as

$$L(p|r) = p^r (1 - p)^{n-r} \quad \text{for observed } r.$$

Suppose $H_0 : p = \frac{1}{2}$, $H_1 : p < \frac{1}{2}$ and the critical region is $s = \{0, 1, 2, \ldots, c\}$, where

$$\text{size} = \alpha = \sum_{j=0}^{c} \binom{n}{j} \left(\frac{1}{2}\right)^n;$$

$$\text{power} = 1 - \beta(p) = \sum_{j=0}^{c} \binom{n}{j} p^j (1 - p)^{n-j}.$$

Consider the following:

$$P[R \in s|p] = 1 - \beta(p) \quad 0 < p \leq \frac{1}{2}.$$

Table 4.5: Size and Power H_0 versus H_2

	Size	Power
T_1	.01	.970
T_2	.01	.020
T_3	.01	.019
T_4	.01	.971

Now suppose the statisticians were given the following probabilities from a less informative binary experiment, where only $R \in s$ or $R \notin s$ is observed:

$$P(R \in s|H_1) = 1 - \beta(p), \qquad P(R \notin s|H_1) = \beta(p),$$
$$P(R \in s|H_0) = \alpha = 1 - \beta(1/2), \qquad P(R \notin s|H_0) = 1 - \alpha = \beta(1/2).$$

Hence

$$L(p|R \in s) = 1 - \beta(p), \qquad L(p|R \notin s) = \beta(p).$$

$$\text{Then } Q = \frac{L(H_0)}{L(H_1)} = \frac{1 - \beta(\frac{1}{2})}{1 - \beta(p)} = \frac{\alpha}{1 - \beta(p)} \quad \text{if} \quad R \in s,$$

$$Q = \frac{L(H_0)}{L(H_1)} = \frac{\beta(\frac{1}{2})}{\beta(p)} = \frac{1 - \alpha}{\beta(p)} \quad \text{if} \quad R \notin s.$$

Note the likelihood for the less informative experiment is exactly the power function or 1 minus the power function from the initially more informative experimental setup. A less pleasant way of saying this is that the N-P approach may not use all of the information available since it changes the situation into a Bernoulli variate by collapsing the sample space.

Example 4.2

Let $\{X_i, i = 1, \ldots, 5\}$ be i.i.d. Bernoulli trials with probability p of success. Let $R = \sum_{i=1}^5 X_i$, $P[X_i = 1] = p = 1 - P[X_i = 0]$.
Consider

$$H_0 : p = \frac{1}{2}, \qquad H_1 : p = 0.1,$$

$$P(R = r|p) = \binom{5}{r} p^r (1 - p)^{5-r}.$$

If we want a test of size exactly $1/16$ the only one available is to reject if $R = 0$ or 5 but to reject H_0 in favor of H_1 when $R = 5$ is absurd. This calls for something else, namely, a randomized test which rejects at $R = 0$ and at $R = 1$ with probability $1/5$ since

$$\epsilon = P\left(R = 0|p = \frac{1}{2}\right) + \frac{1}{5}P\left(R = 1|p = \frac{1}{2}\right) = \frac{1}{32} + \frac{1}{5} \times \frac{5}{32} = \frac{1}{16},$$

which is the required size. It has power $1 - \beta = P(R = 0|p = .1) + \frac{1}{5}P(R = 1|p = .1) = .59049 + \frac{1}{5} \times .32805 = .6561$. This will be the MP test of size $\leq \frac{1}{16}$ that is, of level $\frac{1}{16}$. This is the kind of test advocated by N-P theory.

So for N-P theory we will use the following notation. Let $T(D)$ be the probability of rejecting H_0 so that $0 \leq T(D) \leq 1$ for each D. Now for a continuous $f(D)$

Table 4.6: Binomial Probabilities $p=\frac{1}{2}$ and 0.1

R	5	4	3	2	1	0
$P(R|p=\frac{1}{2})$	$\frac{1}{32}$	$\frac{5}{32}$	$\frac{10}{32}$	$\frac{10}{32}$	$\frac{5}{32}$	$\frac{1}{32}$
$P(R|p=.1)$	0.00001	0.00045	0.00810	0.07290	0.32805	0.59049

generally speaking $T(D) = 0$ or 1. But for a discrete case

$$\epsilon = E_{H_0}[T(D)] = \sum_{D \in S} T(D)P(D|H_0)$$

$$1 - \beta = E_{H_1}[T(D)] = \sum_{D \in S} T(D)P(D|H_1)$$

or if Lebesgue-Stieltjes integrals are used

$$H_0 : F = F_0(D) \quad \text{vs.} \quad H_1 : F = F_1(D);$$

$$\epsilon = \int_S T(D)\,dF_0 \quad \text{and} \quad 1 - \beta = \int_S T(D)\,dF_1.$$

The use of Radon-Nykodym derivatives for the generalized densities depending on a common measure leads to

$$H_0 : f_0 = f_0(D) \quad \text{vs.} \quad H_1 : f_1 = f_1(D)$$

$$P(D \in A|H_0) = \int_A f_0(D)d\mu$$

$$P(D \in A|H_1) = \int_A f_1(D)d\mu \quad \text{for a measurable set A.}$$

The theory of hypothesis testing in its most elegant and general form requires generalized densities (Lehmann, 1959).

4.4 NEYMAN-PEARSON FUNDAMENTAL LEMMA

Lemma 4.1 Let $F_0(D)$ and $F_1(D)$ be distribution functions possessing generalized densities $f_0(D)$ and $f_1(D)$ with respect to μ. Let $H_0 : f_0(D)$ vs. $H_1 : f_1(D)$ and $T(D) = P[\text{rejecting } H_0]$. Then

1. *Existence*: For testing H_0 vs. H_1 there exists a test $T(D)$ and a constant k such that for any given $\alpha, 0 \leq \alpha \leq 1$
   (a) $E_{H_0}(T(D)) = \alpha = \int_S T(D)f_0(D)d\mu$
   (b) $T(D) = \begin{cases} 1 & \text{if} \quad f_0(D) < kf_1(D) \\ 0 & \text{if} \quad f_0(D) > kf_1(D) \end{cases}.$

2. *Sufficiency*: If a test satisfies (a) and (b) then it is the most powerful (MP) for testing H_0 vs. H_1 at level α.

3. *Necessity*: If $T(D)$ is most powerful (*MP*) at level α for H_0 vs. H_1 then for some k it satisfies (a) and (b) unless there exists a test of size less than α and power 1.

Proof: For $\alpha = 0$ or $\alpha = 1$ let $k = 0$ or ∞ respectively, hence we restrict α such that $0 < \alpha < 1$.

1. *Existence*: Let

$$\alpha(k) = P(f_0 \le kf_1|H_0) = P(f_0 < kf_1|H_0) + P(f_0 = kf_1|H_0).$$

Now for $k < 0$, $\alpha(k) = 0$ and $\alpha(k)$ is a monotone non-decreasing function of k continuous on the right and $\alpha(\infty) = 1$. Hence $\alpha(k)$ is a distribution function. Therefore k is such that

$$\alpha(k - 0) \le \alpha \le \alpha(k), \qquad P\left(\frac{f_0}{f_1} < k|H_0\right) \le \alpha \le P\left(\frac{f_0}{f_1} \le k|H_0\right).$$

Hence

$$P\left(\frac{f_0}{f_1} = k|H_0\right) = \alpha(k) - \alpha(k - 0).$$

Now let $T(D)$ be such that

$$T(D) = \begin{cases} 1 & f_0 < kf_1 \\ \dfrac{\alpha - \alpha(k - 0)}{\alpha(k) - \alpha(k - 0)} & f_0 = kf_1 \\ 0 & f_0 > kf_1. \end{cases}$$

If $\alpha(k) = \alpha(k - 0)$ there is no need for the middle term since

$$P(f_0 = kf_1) = 0.$$

Hence we can produce a test with properties (a) and (b).

2. *Sufficiency*: If a test satisfies (a) and (b) then it is most powerful for testing H_0 against H_1 at level α. ☐

Proof: Let $T^*(D)$ be any other test such that

$$E_{H_0}(T^*(D)) \le \alpha.$$

Now consider

$$E_{H_1}(T^*(D)) = \int_S T^*f_1 d\mu = 1 - \beta^*.$$

Then

$$(1 - \beta) - (1 - \beta^*) = \int_S (T - T^*) f_1 d\mu.$$

Divide S into 3 regions such that

S^+ is the set of points D where $T > T^*$, $T > 0$;

S^- is the set of points D where $T < T^*$, $T < 1$;

S^0 is the set of points D where $T = T^*$.

For $D \in S^+$, $T > 0$ and $f_0 \leq kf_1$ or $\frac{f_0}{k} \leq f_1$. For $D \in S^-$, $T < 1$ and $f_0 \geq kf_1$ or $\frac{f_0}{k} \geq f_1$.
Hence

$$(1 - \beta) - (1 - \beta^*)$$

$$= \int_{S^+} (T - T^*) f_1 d\mu + \int_{S^-} (T - T^*) f_1 d\mu$$

$$\geq \int_{S^+} (T - T^*) \frac{f_0}{k} d\mu - \int_{S^-} (T^* - T) f_1 d\mu$$

$$\geq \frac{1}{k} \int_{S^+} (T - T^*) f_0 d\mu - \frac{1}{k} \int_{S^-} (T^* - T) f_0 d\mu$$

$$= \frac{1}{k} \int_{S^+ \cup S^-} (T - T^*) f_0 d\mu = \frac{1}{k} (E_{H_0}(T) - E_{H_0}(T^*))$$

$$= \frac{1}{k} (\alpha - E_{H_0}(T^*)) \geq 0,$$

since $E_{H_0}(T^*) \leq \alpha$. Hence $T(D)$ as defined is most powerful (MP) at level α.

3. *Necessity*: If a test is most powerful at level α then it is defined as in (a) and (b).   □

Proof: Let T^* be MP at level α and let T satisfy (a) and (b). Now $S^+ \cup S^-$ is the set of points D where $T \neq T^*$. Let $S^{\neq}$ be the set of points where $f_0 \neq kf_1$.
   Now $S^{\neq} \cap (S^+ \cup S^-) = s$ is the set of all points D for which either

$$\{T^* < T \quad \text{and} \quad f_0 < kf_1, \}$$

or

$$\{T^* > T \quad \text{and} \quad f_0 > kf_1. \}$$

This can be now represented as follows:

	$f_0 < kf_1$	$f_0 > kf_1$
$T > T^* = S^+$	s	$\emptyset$
$T < T^* = S^-$	$\emptyset$	s

Now $T > T^* \Rightarrow f_0 < kf_1$ and $T < T^* \Rightarrow f_0 > kf_1$. If the measure of s, $\mu(s) > 0$ then

$$(1 - \beta) - (1 - \beta^*) = \int_s (T - T^*)\left(f_1 - \frac{1}{k}f_0\right)d\mu > 0 \quad (A)$$

since the integrand is greater than zero. However, (A) contradicts the assumption that T^* is MP. Therefore $\mu(s) = 0$ *a.e.*

Now suppose $E_{H_0}(T^*) < \alpha$ and $1 - \beta^* < 1$. Then it would be possible to add a set of points, possibly by randomization, to the rejection region until either the size is α or the power is one. Thus either $E_{H_0}(T^*) = \alpha$ or $E_{H_1}(T^*) = 1$. Therefore we have shown that the MP test is uniquely determined by (a) and (b) except on the set of points S^0 where $f_0 = kf_1$ and T will be constant on their boundary $f_0 = kf_1$.

Lastly, if there exists a test of power 1, the constant k in (b) is $k = \infty$ and one will accept H_0 only for those points where $f_0 = kf_1$ even though the size may be less than α. If $P(f_0 = kf_1) = 0$ then the unique test is nonrandomized and is most powerful for testing a simple H_0 against a simple H_1. Note that if there exists a test of power 1 the MP test at level α need not be unique since there might be a test satisfying (a) and (b) with $\alpha' < \alpha$. □

We now show that, in the simple versus simple case, the MP test is also unbiased.

Corollary 4.1 (Unbiasedness). Let $1 - \beta$ denote the power of the MP level α test for $H_0 : f_0$ vs. $H_1 : f_1$. Then $\alpha < 1 - \beta$ unless $F_0 = F_1$.

Proof: Let $T(D)$ be the test that rejects H_0 with probability α no matter what D is. Therefore

$$E_{H_0}(T) = \alpha = E_{H_1}(T), \quad \alpha \leq 1 - \beta.$$

If $\alpha = 1 - \beta$ then T is most powerful and must satisfy (a) and (b) of the N-P lemma, which it contradicts unless $f_0 = f_1$, so $F_0 = F_1$. □

Reversing the inequalities in the NP test results in the least powerful test as will be seen in the next corollary.

Corollary 4.2 (Smallest Power). For $H_0 : f_0$ vs. $H_1 : f_1$, the test

$$T(D) = \begin{cases} 1 & \text{if } f_0 > kf_1 \\ 0 & \text{if } f_0 < kf_1 \end{cases}$$

has the smallest power among all tests with size $\geq \alpha$.

Proof: Let any other test T^* be of size $\geq \alpha$. Then it has power $1 - \beta^*$. Now

$$(1 - \beta^*) - (1 - \beta) = \int_{S^+} (T^* - T)f_1 d\mu + \int_{S^-} (T^* - T)f_1 d\mu,$$

where

$$S^+ = \text{all points } D \text{ such that } T^* < T \text{ so that } T > 0 \text{ and } f_0 \geq kf_1;$$
$$S^- = \text{all points } D \text{ such that } T^* > T \text{ so that } T < 1 \text{ and } f_0 \leq kf_1.$$

Then

$$(1 - \beta^*) - (1 - \beta) \geq \frac{1}{k}\int_{S^+} (T^* - T)f_0 d\mu + \frac{1}{k}\int_{S^-} (T^* - T)f_0 \, d\mu$$

$$\geq \frac{1}{k}\int_{S} (T^* - T)f_0 d\mu = \frac{1}{k}(\alpha^* - \alpha) \geq 0.$$

$\square$

The following corollary establishes that rejection regions for NP tests with smaller size are nested within rejection regions for NP tests with larger size.

Corollary 4.3 (Nesting Corollary) Suppose $T(D)$ satisfies

$$T = \begin{cases} 1 & f_0 < k_\alpha f_1 \\ 0 & f_0 > k_\alpha f_1 \end{cases},$$

with rejection set s_α and

$$T'(D) = \begin{cases} 1 & f_0 < k_{\alpha'} f_1 \\ 0 & f_0 > k_{\alpha'} f_1 \end{cases},$$

with rejection set $s_{\alpha'}$. Then $\alpha' > \alpha$ implies $s_\alpha \subseteq s_{\alpha'}$.

Proof: Now assume $k_{\alpha'} < k_\alpha$ so that

$$\alpha = P\left(\frac{f_0}{f_1} < k_\alpha | H_0\right) + pP\left(\frac{f_0}{f_1} = k_\alpha | H_0\right) \equiv a + pb$$

$$< P\left(\frac{f_0}{f_1} < k_{\alpha'} | H_0\right) + p'P\left(\frac{f_0}{f_1} = k_{\alpha'} | H_0\right) \equiv a' + p'b' = \alpha'.$$

But this implies that

$$a + pb < a' + p'b' \leq a' + b'.$$

Since $a \geq a'$, this implies that $p'b' - pb > a - a' \geq 0$. But

$$a - a' = b' + P\left(k_{\alpha'} < \frac{f_0}{f_1} < k_{\alpha}\right),$$

so that

$$0 \geq b'(p' - 1) - pb > P\left(k_{\alpha'} < \frac{f_0}{f_1} < k_{\alpha}\right) \geq 0,$$

which is a contradiction. Hence $k_{\alpha'} \geq k_{\alpha}$ and $s_{\alpha} \subseteq s_{\alpha'}$.   □

Therefore it has been shown that the MP level α tests for simple H_0 vs. simple H_1 are nested in that for $\alpha < \alpha'$, $s_{\alpha} \subset s_{\alpha'}$. But this need not hold for a composite null H. If $H_i \in H$ and $K_j \in K$, then if among all tests T^* of level α, that is,

$$E(T^*|H) \leq \alpha \quad \text{for all } H_i \in H,$$

there exists a test T such that

$$E(T^*|K_j) \leq E(T|K_j) \quad \text{for all } K_j \in K,$$

then the test based on T is uniformly most powerful (UMP). Often $H : \theta \in \Theta_H$ vs. $K : \theta \in \Theta_K$. When UMP tests exist they are the ones recommended by N-P theory. In the case of a simple H_0 versus composite H_1 these UMP tests are also likelihood tests under fairly general conditions, but UMP tests do not exist in many situations and not all likelihood tests are UMP.

The next lemma asserts that UMP tests are unbiased.

Lemma 4.2   If a UMP test of level α exists and has power $1 - \beta_K$, then $\alpha \leq 1 - \beta_K$, for all alternatives K.

Proof: Since an α level test $T^*(D) = \alpha$ exists

$$E(T^*(D)\,|H) \equiv \alpha,$$

and

$$E(T^*(D)\,|K) \equiv \alpha = 1 - \beta_K^*.$$

Assume T is of level α and UMP. Then the power of T is at least as great as $1 - \beta_K^*$ so that $\alpha = 1 - \beta_K^* \leq 1 - \beta_K$.   □

Consider the following counterexample (Stein, 1951) to the nesting corollary when H is composite.

Table 4.7: Probabilities under H and K

D		1	2	3	4
H	H_0	$\frac{2}{13}$	$\frac{4}{13}$	$\frac{3}{13}$	$\frac{4}{13}$
	H_1	$\frac{4}{13}$	$\frac{2}{13}$	$\frac{1}{13}$	$\frac{6}{13}$
K	H_2	$\frac{4}{13}$	$\frac{3}{13}$	$\frac{2}{13}$	$\frac{4}{13}$

Example 4.3

Table 4.7 presents a scenario for $H : H_0$ or H_1 vs. $K : H_2$. Consider testing H vs. K of level $\frac{5}{13}$. Consider the test which rejects H_0 or H_1 when $D = 1$ or 3. This test has size $\frac{5}{13}$ and power $\frac{6}{13}$. It is easy to show that this is UMP for rejecting H vs. K at level $\frac{5}{13}$.

Now consider the test which rejects H_0 or H_1 at level $\frac{6}{13}$. Note if $D = 1$ or 2 we reject at level $\frac{6}{13}$ and this test has power $\frac{7}{13}$. One can also show that this is UMP at level $\frac{6}{13}$. Hence if $D = 3$ we reject at level $\frac{5}{13}$ but not at $\frac{6}{13}$. One then cannot really argue that in all cases the level of rejection measures how strongly we believe the data contradict the hypotheses. We also note that these tests are not likelihood tests. Recall the definition of a likelihood test for H vs. K:

Reject if for every $H_i \in H$ and some $K_j \in K$,

$$Q(D) = \frac{P(D|H_i)}{P(D|K_j)} < a.$$

In our case reject if

$$Q(D) = \frac{\max\ P(D|H_i)}{P(D|K)} < a.$$

Using Table 4.8, possible rejection regions are:

a	R
$a > \frac{3}{2}$	$\{1, 2, 3, 4\}$
$\frac{4}{3} < a \leq \frac{3}{2}$	$\{1, 2\}$
$1 < a \leq \frac{4}{3}$	$\{1\}$
$a \leq 1$	$\{\emptyset\}$

Notice that $D = \{1, 3\}$ can never be the rejection region of the Likelihood test. So the UMP test is not a likelihood test.

Table 4.8: Likelihood Ratios

D	1	2	3	4
$Q(D)$	1	$\frac{4}{3}$	$\frac{3}{2}$	$\frac{3}{2}$

4.5 MONOTONE LIKELIHOOD RATIO PROPERTY

For scalar θ we suppose $f(\theta|D) = L(\theta|D) = L(\theta|t(D))$ and for all $\theta' < \theta''$,

$$Q(t(D)|\theta', \theta'') = \frac{L(\theta'|t(D))}{L(\theta''|t(D))}$$

is a non-decreasing function of $t(D)$ or as $t(D) \uparrow Q \uparrow$, that is, the support for θ' vs. θ'' does not decrease as t increases and does not increase as t decreases. We call this a monotone likelihood ratio (MLR) property of $L(\theta|t(D))$.

Theorem 4.1 Let $f(D|\theta)$ have the MLR property. To test $H : \theta \le \theta_0$ vs. $K : \theta > \theta_0$ the following test $T(D)$ is UMP at level α.

$$T(D) = \begin{cases} 1 & \text{if} \quad t(D) < k \quad \text{e.g., reject } H \text{ at level } \alpha, \\ p & \text{if} \quad t(D) = k \quad \text{e.g., reject } H \text{ with probability } p, \\ 0 & \text{if} \quad t(D) > k \quad \text{e.g., accept } H, \end{cases}$$

where k and p are determined by

$$\alpha = E_{\theta_0}(T(D)) = P(t(D) < k|\theta_0) + pP(t(D) = k|\theta_0)$$

with power

$$1 - \beta(\theta) = E_\theta(T(D)) = P(t(D) < k|\theta) + pP(t(D) = k|\theta),$$

which is an increasing function of θ for all values θ such that $1 - \beta(\theta) < 1$ and for any $\theta < \theta_0$ the test minimizes $1 - \beta(\theta)$ among all test satisfying

$$E_{\theta_0}(T(D)) \ge \alpha.$$

Proof: To show the test exists and $E_{\theta_0}(T) = \alpha$, $E_\theta(T) = 1 - \beta(\theta)$ apply the NP lemma to $H : \theta = \theta_0$ vs. $K : \theta = \theta_1 > \theta_0$ so there exists a k and a p such that the test satisfies (a) and (b), that is,

$$Q = \frac{f(D|\theta_0)}{f(D|\theta_1)} < c \text{ implies } t(D) < k \text{ since } Q \uparrow \quad \text{for } t(D) \uparrow .$$

This test is MP for testing

$$H : \theta = \theta' \text{ vs. } K : \theta = \theta'' > \theta' \text{ at level } \alpha' = 1 - \beta(\theta').$$

Since $\alpha' < 1 - \beta(\theta'')$ by the unbiasedness Corollary 4.1 then for all $\theta' < \theta''$, $1 - \beta(\theta') < 1 - \beta(\theta'')$ and $1 - \beta(\theta)$ is an increasing function for all θ such that $1 - \beta(\theta) < 1$. Since $1 - \beta(\theta)$ is nondecreasing the test satisfies:

(i) $1 - \beta(\theta) = E_\theta(T(D)) \le \alpha = 1 - \beta(\theta_0)$ for $\theta \le \theta_0$.
    Further the class of tests satisfying (i) is contained in the class satisfying

(ii) $E_{\theta_0}(T(D)) \le \alpha$ (less restrictions in (ii) than in (i)).
    Since the given test maximizes $1 - \beta(\theta_1)$ within the wider class (ii) it also maximizes $1 - \beta(\theta_1)$ subject to (i) and since it does not depend on the particular alternative $\theta_1 > \theta_0$ it is UMP against alternative $K : \theta > \theta_0$ when $H : \theta \le \theta_0$.   □

To show that $1 - \beta(\theta)$ is minimized for $\theta < \theta_0$, we note that this test reverses the inequalities of the UMP test of $\theta = \theta_0$ versus $\theta = \theta_1 < \theta_0$. Hence it has minimum power by the smallest power corollary for any test of size $\ge \alpha$. Note that if $Q \uparrow$ as $t(D) \downarrow$, the inequalities are reversed:

$$Q < c \implies \begin{array}{ll} t(D) > k & \text{reject } H \\ t(D) = k & \text{reject } H \text{ with probability } p \\ t(D) < k & \text{accept } H \end{array}$$

and the definition of $1 - \beta(\theta)$ is

$$1 - \beta(\theta) = P(t(D) > k | \theta) + pP(t(D) = k).$$

Further, we can turn the hypotheses around

$$H : \theta \ge \theta_0 \quad K : \theta < \theta_0.$$

Then if $Q \uparrow$ as $t(D) \uparrow$ then $Q < c \Rightarrow t(D) < k$ and

$$\alpha = P(t(D) < k | \theta_0) + pP(t(D) = k)$$
$$1 - \beta(\theta) = P(t(D) < k | \theta) + pP(t(D) = k).$$

Corollary 4.4 Let θ be a scalar parameter and

$$f(D|\theta) = c(\theta)e^{q(\theta)t(D)}h(D),$$

where the range of D does not depend on θ and $q(\theta)$ is strictly monotonic in θ. Then there exists a UMP test $T(D)$ for $H : \theta \le \theta_0$ vs. $K : \theta > \theta_0$.

Proof: Note that

$$Q = \frac{f(D|\theta')}{f(D|\theta'')} = \frac{c(\theta')}{c(\theta'')}e^{t(D)[q(\theta')-q(\theta'')]}$$

for $\theta' < \theta''$. Now as $t(D)$ increases Q either increases or decreases depending on whether $q(\theta)$ decreases or increases, that is,

$$\text{if } q(\theta) \downarrow \text{ then } q(\theta') > q(\theta'') \text{ and } Q \uparrow \text{ as } t \uparrow$$
$$\text{if } q(\theta) \uparrow \text{ then } Q \uparrow \text{ as } t \downarrow.$$

Therefore, the monotone likelihood ratio property holds and the corollary is established.

4.6 DECISION THEORY

Suppose we now consider the testing problem from a decision theoretic point of view, that is, testing $H : \theta \leq \theta_0$ vs. $K : \theta > \theta_0$ with d_0 the decision to accept H and d_1 the decision to accept K. Assume a loss function $l(\theta, d_i)$. Now it appears to be sensible to assume

$$l(\theta, d_0) = 0 \quad \text{if} \quad \theta \leq \theta_0 \text{ and } \uparrow \text{ for } \theta > \theta_0,$$
$$l(\theta, d_1) = 0 \quad \text{if} \quad \theta > \theta_0, \text{and } \uparrow \text{ for } \theta \leq \theta_0.$$

Hence

(1) $l(\theta, d_1) - l(\theta, d0) \begin{cases} > 0 & \text{for } \theta \leq \theta_0 \\ < 0 & \text{for } \theta > \theta_0. \end{cases}$

Now recall that the risk function of a test T for a given θ is the average loss

$$l(\theta, T) = T(D)l(\theta, d_1) + (1 - T(D))l(\theta, d_0),$$
$$R(\theta, T) = E_D l(\theta, T) = E_D[T(D)l(\theta, d_1) + (1 - T)l(\theta, d_0)].$$

Further suppose for tests $T(D)$ and $T^*(D)$ that

(2) $R(\theta, T^*) \leq R(\theta, T)$ for all θ,

(3) $R(\theta, T^*) < R(\theta, T)$ for some θ.

Then T^* dominates T and we say T is inadmissible. On the other hand if no such T^* exists then T is admissible. A class C of test (decision) procedures is complete if for any test T not in C there exists a test T^* in C dominating it. (A complete class is minimal if it does not contain a complete subclass). A class C is essentially complete if at least (2) holds. (A complete class is also essentially complete, a class is minimal essentially complete if it does not contain an essentially complete subclass).

Theorem 4.2 Under the assumption of the monotone likelihood ratio theorem, the family of tests given there with $0 \leq \alpha \leq 1$ is essentially complete for the loss function given by (1). Also the family is minimal essentially complete provided the set of points D for which $f(D|\theta) > 0$ is independent of θ.

Proof: The risk of any test T is

$$R(\theta, T) = \int f(D|\theta)[Tl(\theta, d_1) + (1 - T)l(\theta, d_0)]d\mu$$

$$= \int f(D|\theta)[l(\theta, d_0) + T(l(\theta, d_1) - l(\theta, d_0)]d\mu.$$

Similarly for a UMP test T^*,

$$R(\theta, T^*) - R(\theta, T) = \int f(D|\theta)(T^* - T)(l(\theta, d_1) - l(\theta, d_0))d\mu$$

$$= (l(\theta, d_1) - l(\theta, d_0))E_\theta(T^* - T) \le 0,$$

since (i) $l(\theta, d_1) - l(\theta, d_0) > 0$, and $E_\theta(T^* - T) \le 0$ for $\theta < \theta_0$, which follows because T^* is UMP so it minimizes power over H, and since (ii) $l(\theta, d_1) - l(\theta, d_0) < 0$ and $E_\theta(T^* - T) \ge 0$ for $\theta > \theta_0$. Therefore $R(\theta, T^*) \le R(\theta, T)$ for all θ. Thus the family T^* is essentially complete since (2) holds.

To show that the class is minimal essentially complete choose two tests T_1^* and T_2^* that belong to class UMP and have sizes α_1 and α_2 for testing $\theta = \theta_0$ vs. $\theta > \theta_0$. Then for $\alpha_1 < \alpha_2$,

$$E_\theta(T_1^*) < E_\theta(T_2^*) \; \forall \, \theta > \theta_0,$$

since T_2^* is UMP at level α_2, unless $E_\theta(T_1^*) = 1$. Further,

$$E_\theta(T_1^*) < E_\theta(T_2^*) \; \forall \; \theta < \theta_0$$

because at level $\alpha_1 < \alpha_2$, T_1^* has minimum power unless $E_\theta(T_2^*) = 0$ for $\theta = \theta_0$. Now since

$$R(\theta, T_1^*) - R(\theta, T_2^*) = [l(\theta, d_1) - l(\theta, d_0)]E_\theta(T_1^* - T_2^*),$$
$$R(\theta, T_1^*) > R(\theta, T_2^*) \; \text{if} \; \theta > \theta_0, \text{and}$$
$$R(\theta, T_1^*) < R(\theta, T_2^*) \; \text{if} \; \theta < \theta_0.$$

Since $E_\theta(T_2^*) = 0$ and $E_\theta(T_1^*) = 1$ are excluded by assumption each of the two risk functions are better than the other for some values of θ. Hence the family is minimal essentially complete. ☐

4.7   TWO-SIDED TESTS

For two-sided tests we have, in particular,

$$H : \theta = \theta_0 \quad \text{vs.} \quad K : \theta \neq \theta_0.$$

Do UMP tests exist here? Yes, but only in unusual cases. Note for the MLR property one-sided UMP tests exist so for these situations the answer is no!

Example 4.4

Consider the example $X_1, \ldots, X_n$ i.i.d. from

$$f(x) = \begin{cases} e^{-(x-\theta)} & \text{for } \theta \leq x \\ 0 & \text{elsewhere.} \end{cases}$$

To test $H_0 : \theta = \theta_0$ vs. $H_1 : \theta \neq \theta_0$ we first look at $H_0 : \theta = \theta_0$ vs. $H_1 : \theta = \theta_1 > \theta_0$. Let $x_{(1)}$ be the smallest observation. Then

$$L(\theta | x_1, \ldots, x_n) = \begin{cases} e^{-\Sigma(x_i-\theta)} = e^{-n\bar{x}+n\theta} & x_{(1)} \geq \theta \\ 0 & x_{(1)} < \theta \end{cases},$$

and

$$Q = \frac{L(\theta_0)}{L(\theta_1)} = \begin{cases} \infty & \theta_0 \leq x_{(1)} < \theta_1 \\ e^{n(\theta_0-\theta_1)} < 1 & \theta_0 < \theta_1 \leq x_{(1)}. \end{cases}$$

For the one-sided alternative the MLR property holds based on $x_{(1)}$, that is, $Q \downarrow$ as $x_{(1)} \uparrow$ so the UMP leavel α test for $H : \theta = \theta_0$ versus $K : \theta > \theta_0$ rejects for $x_{(1)} > c_\alpha$. Similarly, the UMP test for $H : \theta = \theta_0$ versus $K : \theta < \theta_0$ rejects for $x_{(1)} < c_\alpha^*$.
  The density of $X_{(1)}$ is

$$f(x_{(1)} | \theta) = n e^{n(\theta - x_{(1)})} \quad \text{for } x_{(1)} \geq \theta,$$

and thus

$$P_\theta(x_{(1)} > c) = e^{n(\theta-c)}, \quad c > \theta.$$

Then $c_\alpha = \theta_0 - \log(\alpha)/n$ and $c_\alpha^* = \theta_0 - \log(1-\alpha)/n$. The corresponding powers are:

$$\beta(\theta) = \begin{cases} \alpha e^{n(\theta-\theta_0)} & \theta \leq c_\alpha, \\ 1 & \theta > c_\alpha \end{cases} \quad \theta \geq \theta_0,$$

$$\beta^*(\theta) = 1 - (1-\alpha)e^{n(\theta-\theta_0)}, \quad \theta \leq \theta_0.$$

Now for $H_0 : \theta = \theta_0$ vs. $H_1 : \theta \neq \theta_0$,

$$Q = \frac{L(\theta_0)}{L(\theta_1)} = \begin{cases} \infty & \text{if} \quad \left.\begin{matrix} \theta_0 \leq x_{(1)} < \theta_1 \\ \theta_0 < \theta_1 \leq x_{(1)} \end{matrix}\right\}Q \uparrow \quad \text{as} \ x_{(1)} \downarrow \\ e^{n(\theta_0 - \theta_1)} < 1 \\ e^{n(\theta_0 - \theta_1)} > 1 & \left.\begin{matrix} \theta_1 < \theta_0 \leq x_{(1)} \\ \theta_1 \leq x_{(1)} < \theta_0 \end{matrix}\right\}Q \uparrow \quad \text{as} \ x_{(1)} \uparrow \\ 0 \end{cases}.$$

The MLR property does not hold, but consider the rejection region $\{x_{(1)} < \theta_0$ or $x_{(1)} > c_\alpha\}, P(x_{(1)} < \theta_0|H_0) = 0$ and $P(x_{(1)} > c_\alpha|H_0) = \alpha$. But it is easy to show that this test has the same power as the above two tests in their respective alternatives. So the test is UMP for $\theta_0 > \theta_1$ and also for $\theta_0 < \theta_1$.

We now show that for testing $\theta = \theta_0$ vs. $\theta \neq \theta_0$ a necessary condition for a UMP test to exist is $d \log L(\theta|D)/d\theta = \text{const} \neq 0$. Assuming $d \log L(\theta|D)/d\theta$ is continuous with respect to θ, for any $\theta_1 \neq \theta_0$ expand in a Taylor series $L(\theta_1|D) = L(\theta_0|D) + (\theta_1 - \theta_0)L'(\theta^*|D)$ where $\theta^* \in (\theta_0, \theta_1)$. Now if a UMP test exists then the rejection region for any particular θ_1 is

(a)
$$\frac{L(\theta_1|D)}{L(\theta_0|D)} = 1 + \frac{(\theta_1 - \theta_0)L'(\theta^*|D)}{L(\theta_0|D)} \geq k_\alpha(\theta_1)$$

(and randomize on the boundary if necessary). For all D,

$$k_\alpha(\theta_0) = 1 = \frac{L(\theta_0|D)}{L(\theta_0|D)}$$

so

(b) $k_\alpha(\theta_1) = 1 + (\theta_1 - \theta_0)k'_\alpha(\theta^{**})$ where $\theta^{**} \in (\theta_0, \theta_1)$.
Substitute (b) in (a) and obtain
(c) $(\theta_1 - \theta_0)\left\{\frac{L'(\theta^*|D)}{L(\theta_0|D)} - k'_\alpha(\theta^{**})\right\} \geq 0.$

For the points d on the boundary of the rejection region,

$$1 + \frac{(\theta_1 - \theta_0)L'(\theta^*|d)}{L(\theta_0|d)} = 1 + (\theta_1 - \theta_0)k'_\alpha(\theta^{**})$$

or

$$k'_\alpha(\theta^{**}) = \frac{L'(\theta^*|d)}{L(\theta_0|d)}$$

and from (c),

$$(\theta_1 - \theta_0)\left\{\frac{L'(\theta^*|D)}{L(\theta_0|D)} - \frac{L'(\theta^*|d)}{L(\theta_0|d)}\right\} \geq 0.$$

This holds for all θ_1, and all D in the rejection region including d. Now since $\theta_1 - \theta_0$ changes sign the quantity in the brackets must be equal to 0.

For the points D outside of the rejection region $\frac{L(\theta_1|D)}{L(\theta_0|D)} \leq k_\alpha(\theta_1)$ so the same argument leads to

$$(\theta_1 - \theta_0)\left\{\frac{L'(\theta^*|D)}{L(\theta_0|D)} - \frac{L'(\theta^*|d)}{L(\theta_0|d)}\right\} \leq 0$$

for all D not in the rejection region but including d. Hence again the quantity in the brackets must be equal—this is true for all D inside or outside the rejection region. Therefore,

$$\frac{L'(\theta^*|D)}{L(\theta_0|D)} = \frac{L'(\theta^*|d)}{L(\theta_0|d)} = \text{constant}.$$

Then the assumed continuity of L' with respect to θ implies that as $\theta_1 \to \theta_0$

$$\frac{L'(\theta^*|D)}{L(\theta_0|D)} = \frac{L'(\theta_0|D)}{L(\theta_0|D)} = \frac{d\log L(\theta|D)}{d\theta}\bigg|_{\theta=\theta_0} = \text{constant (in } D).$$

Since the proof does not depend on a particular θ_0 it holds for all θ. Note if

$$E\left[\frac{d\log L(\theta|D)}{d\theta}\right] = 0,$$

then

$$\frac{d\log L(\theta|D)}{d\theta} \neq \text{constant}.$$

REFERENCES

Hacking, I. (1965). *Logic of Statistical Inference*. Cambridge: Cambridge University Press.

Lehmann, E.L. (1959). *Testing Statistical Hypothesis*. New York: Wiley.

Neyman, J. and Pearson, E. S. (1933). On the problem of the most efficient test of statistical hypothesis. *Philosophical Transactions of the Royal Society* **LCXXXI**, 289–337.

CHAPTER FIVE

Unbiased and Invariant Tests

Since UMP tests don't always exist, statisticians have proceeded to find optimal tests in more restricted classes. One such restriction is unbiasedness. Another is invariance. This chapter develops the theory of uniformly most powerful unbiased (UMPU) and invariant (UMPI) tests. When it is not possible to optimize in these ways, it is still possible to make progress, at least on the mathematical front. Locally most powerful (LMP) tests are those that have greatest power in a neighborhood of the null, and locally most powerful unbiased (LMPU) tests are most powerful in a neighborhood of the null, among unbiased tests. These concepts and main results related to them are presented here. The theory is illustrated with examples and moreover, examples are given that illustrate potential flaws. The concept of a "worse than useless" test is illustrated using a commonly accepted procedure. The sequential probability ratio test is also presented.

5.1 UNBIASED TESTS

Whenever a UMP test exists at level α, we have shown that $\alpha \leq 1 - \beta_K$, that is, the power is at least as large as the size. If this were not so then there would be a test $T^* \equiv \alpha$ which did not use the data and had greater power. Further, such a test is sometimes termed as "worse than useless" since it has smaller power than the useless test $T^* \equiv \alpha$.

Now UMP tests do not always exist except in fairly restricted situations. Therefore N-P theory proposes that in the absence of a UMP test, a test should at least be unbiased, that is, $1 - \beta_K \geq \alpha$. Further, if among all unbiased tests at level α, there exists one that is UMP then this is to be favored and termed UMPU. So for a class of problems for which UMP tests do not exist there may exist UMPU tests.

Modes of Parametric Statistical Inference, by Seymour Geisser
Copyright © 2006 John Wiley & Sons, Inc.

5.2 ADMISSIBILITY AND TESTS SIMILAR ON THE BOUNDARY

A test T at level α is called admissible if there exists no other test that has power no less than T for all alternatives while actually exceeding it for some alternatives, that is, no other test has power that dominates the power of test T. Hence if T is UMPU then it is admissible, that is, if a test T^* dominates T then it also must be unbiased and hence T is not UMPU which contradicts the assumption of T being UMPU. Hence T is admissible.

Recall a test T which satisfies

$$1 - \beta_T(\theta) \leq \alpha \quad \text{for } \theta \in \Theta_H$$
$$1 - \beta_T(\theta) \geq \alpha \quad \text{for } \theta \in \Theta_K$$

is an unbiased test. If T is unbiased and $1 - \beta_T(\theta)$ is a continuous function of θ there is a boundary of values $\theta \in w$ for which $1 - \beta_T(\theta) = \alpha$. Such a level α test with boundary w is said to be similar on the boundary.

Theorem 5.1   If $f(D|\theta)$ is such that for every test T at level α the power function $1 - \beta_T(\theta)$ is continuous and a test T' is UMP among all level α tests similar on the boundary then T' is UMPU.

Proof: The class of tests similar on the boundary contains the class of unbiased tests, that is, if $1 - \beta_T(\theta)$ is continuous then every unbiased test is similar on the boundary. Further if T' is UMP among similar tests it is at least as powerful as any unbiased test and hence as $T^* \equiv \alpha$ so that it is unbiased and hence UMPU.
$\square$

Finding a UMP similar test is often easier than finding a UMPU test directly so that this may provide a way of finding UMPU tests.

Theorem 5.2   Consider the exponential class for a scalar θ

$$f(D|\theta) = C(\theta)e^{\theta t(D)}k(D)$$

with monotonic likelihood ratio property and tests of $H_0 : \theta = \theta_0$ vs. $H_1 : \theta \neq \theta_0$ and $H : \theta_0 \leq \theta \leq \theta_1$ vs. $K : \theta \notin [\theta_0, \theta_1]$. For tests whose power is $\geq \alpha$ we can find one such that it is UMP among all unbiased tests. This test $T(D)$ is such that

$$T(D) = \begin{cases} 1 & \text{if} \quad t(D) > k_1 \quad \text{or } t(D) < k_2 \text{ for } k_1 > k_2 \\ p_1 & \text{if} \quad t(D) = k_1 \\ p_2 & \text{if} \quad t(D) = k_2 \\ 0 & \text{if} \quad k_2 < t(D) < k_1 \end{cases}$$

where k_1, k_2, p_1, and p_2 are determined from:

(a) For $i = 0, 1, E_{\theta_i}(T(D)) = \alpha = P[t > k_1|\theta_i] + p_1 P(t = k_1|\theta_i) + p_2 P(t = k_2|\theta_i) + P(t < k_2|\theta_i)$ and constants k_1, k_2, p_1, p_2 are determined by the above equation.

If $\theta_0 = \theta_1$ so that $H_0 : \theta = \theta_0$ vs. $H_1 : \theta \neq \theta_0$, then $k_1, k_2, p_1,$ and p_2 are determined by

(b) $E_{\theta_0}[t(D)T(D)] = \alpha E_{\theta_0}(t(D))$ assuming it exists that is,

$$\int_{t>k_1} tf(D|\theta_0)d\mu + k_1 p_1 P(t = k_1|\theta_0) + k_2 p_2 P(t = k_2|\theta_0)$$

$$+ \int_{t<k_2} tf(D|\theta_0)d\mu = \alpha \int_S tf(D|\theta_0)d\mu.$$

A proof is given by Lehmann (1959). In particular for $H_0 : \theta = \theta_0$ and $f(t|\theta_0)$ symmetric about a point γ so that $P(t < \gamma - x|\theta_0) = P(t > \gamma + x|\theta_0)$ for all x or $P(t - \gamma < x|\theta_0) = P(t - \gamma > x|\theta_0)$ we can show that $p_1 = p_2, k_1 = 2\gamma - k_2, P[t \leq k_2|\theta_0] = \alpha/2 = P(t| \geq k_1|\theta_0)$ and (b) becomes

$$E_{\theta_0}(tT) = E_{\theta_0}[(t - \gamma)T] + \gamma E_{\theta_0}(T) = E_{\theta_0}[(t - \gamma)T] + \gamma\alpha = \alpha E_{\theta_0}(t) = \alpha\gamma$$

or $E_{\theta_0}[(t - \gamma)T] = 0$.
  It is easy to show that this leads to

$$k_1 - \gamma = \gamma - k_2 \quad \text{and} \quad p_1 = p_2.$$

Consider $H : \theta \in \Theta_H$ vs. $K : \theta \in \Theta_K$
  Recall that a test T is unbiased if

$$1 - \beta_T(\theta) \leq \alpha \text{ for } \theta \in \Theta_H, \qquad 1 - \beta_T(\theta) \geq \alpha \text{ for } \theta \in \Theta_K$$

and if $1 - \beta_T(\theta)$ is continuous there is a boundary of values w such that $1 - \beta_T(\theta) = \alpha$ for $\theta \in w$ and such a level α test is said to be similar on the boundary. If T is a non-randomized test with rejection region s_α

$$P(D \in s_\alpha|\theta \in w) = \alpha \text{ independent of } \theta$$

and, of course, with sample space S

$$P(D \in S|\theta \in \Theta) = 1 \text{ independent of } \theta$$

so s_α is called a similar region, that is, similar to the sample space. Also $t(D)$ will turn out to be a sufficient statistic if $f(D|t)$ is independent of θ (see Section 7.7 for a more formal definition of sufficiency).

5.3 NEYMAN STRUCTURE AND COMPLETENESS

Now if there is a sufficient statistic $t(D)$ for θ then any test T such that

$$E_{D|t}(T(D)|t(D); \ \theta \in w) = \alpha$$

is called a test of Neyman structure (NS) with respect to $t(D)$. This is a level α test since

$$E_{\theta}(T(D)|\theta \in w) = E_t E_{D|t}(T(D)|t(D); \ \theta \in w) = E_t(\alpha) = \alpha.$$

Now it is often easy to obtain a MP test among tests having Neyman structure termed UMPNS. If every similar test has Neyman structure, this test would be MP among similar tests or UMPS. A sufficient condition for a similar test to have Neyman structure is bounded completeness of the family $f(t|\theta)$ of sufficient statistics.

A family $\mathcal{F}$ of probability functions is complete if for any $g(x)$ satisfying

$$E(g(x)) = 0 \quad \text{for all} \quad f \in \mathcal{F},$$

then $g(x) = 0$.

A slightly weaker condition is "boundedly complete" in which the above holds for all bounded functions g.

Theorem 5.3 Suppose for $f(D|\theta)$, $t(D)$ is a sufficient statistic. Then a necessary and sufficient condition for all similar tests to have NS with respect to $t(D)$ is that $f(t)$ be boundedly complete.

Proof of bounded completeness implies NS given similar tests:
Assume $f(t)$ is boundedly complete and let $T(D)$ be similar. Now $E(T(D) - \alpha) = 0$ for $f(D|\theta)$ and if $E(T(D) - \alpha|t(D) = t, \theta \in w) = g(t)$ such that

$$0 = E_D(T(D) - \alpha) = E_t E_{D|t}(T(D) - \alpha)|t(D) = t, \ \theta \in w) = E_t g(t) = 0.$$

Since $T(D) - \alpha$ is bounded then so is $g(t)$ and from the bounded completeness of $f(t)$, $E_t(g(t)) = 0$ implies $g(t) = 0$ such that

$$E(T(D) - \alpha|t(D) = t, \ \theta \in w) = 0 \quad \text{or} \quad E(T(D)|t(D) = t, \ \theta \in w) = \alpha \text{ or NS}.$$

Proof that lack of bounded completeness implies lack of NS:
If $f(t)$ is not boundedly complete then although $g(t)$ is bounded that is, $|g(t)| \leq K$, then $E(g(t)) = 0$ but $g(t) \neq 0$ with positive probability for $f(t)$. Let similar test $T(t) = ag(t) + \alpha$ where $a = \min \frac{1}{K}(\alpha, 1 - \alpha)$ then $T(t)$ is a test function since

$0 \leq T(t) \leq 1$ and it is a similar test since $E(T(t)) = aE(g(t)) + \alpha = \alpha$ for $f(t)$ but it does not have NS since $T(t) \neq \alpha$ with positive probability for some $f(t)$.

Similar tests are desirable when nuisance parameters exist that is, $f(D|\theta, \lambda)$ where $\theta \in \Theta$ and $\lambda \in \Lambda$. Suppose $H_0 : \theta = \theta_0, \lambda \in \Lambda$ vs. $H_1 : \theta \neq \theta_0, \lambda \in \Lambda$ and $f(D|\theta, \lambda)$ admits a sufficient statistic t_λ for every fixed θ in the sense

$$f(D|\theta, \lambda) = f(D|t_\lambda, \theta)f(t_\lambda|\lambda, \theta).$$

Now for $\theta = \theta_0$

$$\alpha = E[T(D)|t_\lambda, \theta_0] = \int T(D)f(D|t_\lambda, \theta_0)d\mu$$

independent of λ. Thus

$$\alpha = E_{t_\lambda}(\alpha) = E_{t_\lambda}E_{D|t_\lambda}[T(D)|t_\lambda, \theta_0] = E_D[T(D)|\lambda, \theta_0]$$

and we have a similar test independent of λ. One can then attempt to find a most powerful test among all similar tests by using the conditional densities.

$$Q = \frac{f_{D|t_\lambda}(D|t_\lambda, \theta_0)}{f_{D|t_\lambda}(D|t_\lambda, \theta_1)} \leq c_\alpha.$$

If the same region is obtained for all $\theta_1 \neq \theta_0, \lambda \in \Lambda$ then we have a level α UMPS test. When this doesn't exist we sometimes can find a UMPUS test, that is, uniformly most powerful unbiased similar test and under certain circumstances this will turn out to be UMPU.

Theorem 5.4 If $f(D|\theta) = C(\theta, \lambda)e^{\theta u(D) + \lambda t(D)}h(D)$ then the UMPU test for $H_0 : \theta = \theta_0, \lambda \in \Lambda$ vs. $H_1 : \theta \neq \theta_0, \lambda \in \Lambda$

$$T(u, t) = \begin{cases} 1 & \text{if} \quad u < k_1(t) \text{ or } u > k_2(t) \\ p_1 & \text{if} \quad u = k_1(t) \\ p_2 & \text{if} \quad u = k_2(t) \\ 0 & \text{if} \quad k_1(t) < u < k_2(t) \end{cases}$$

where k_1, k_2, p_1, and p_2 are determined by

(a) $E_{u|t}[T(u, t)|t, \theta_0] = \alpha$,
(b) $E_{u|t}[uT(u, t)|t, \theta_0] = \alpha E_{u|t}(u|t, \theta_0)$,

and power $1 - \beta(\theta, t) = E_{u|t}[T(u, t)|t, \theta]$.

Note that $E_t(1 - \beta(\theta, t)) = E_t E_{u|t}[T(u, t)|t, \theta] = E_{t, u}(T(u, t)|\theta, \lambda) = 1 - \beta(\theta, \lambda)$ that is, $1 - \beta(\theta, t)$ is an unbiased estimate of $1 - \beta(\theta, \lambda)$ but it cannot be used to determine sample size in advance since $1 - \beta(\theta, t)$ depends on the sample value of t.

For proof, see Lehmann (1959).

Example 5.1

Let

$$f = f(r_1, r_2|p_1, p_2) = \binom{n_1}{r_1}\binom{n_2}{r_2}\prod_{i=1}^{2}p_i^{r_i}(1-p_i)^{n_i-r_i}.$$

For $\lambda = p_2$ and $\theta = p_1(1-p_2)/p_2(1-p_1)$,

$$f = \binom{n_1}{r_1},\binom{n_2}{r_2}\theta^{r_1}\left(\frac{1}{1+\theta(\lambda/1-\lambda)}\right)^{n_1}\left(\frac{\lambda}{1-\lambda}\right)^{r_1+r_2}(1-\lambda)^{n_2}$$

$$= (1-\lambda)^{n_2}\left(\frac{1}{1+\theta(\lambda/1-\lambda)}\right)^{n_1}\theta^{r_1}\left(\frac{\lambda}{1-\lambda}\right)^{r_1+r_2}\binom{n_1}{r_1}\binom{n_2}{r_2}$$

$$= c(\theta,\lambda)e^{u\log\theta+t\log\left(\frac{\lambda}{1-\lambda}\right)}\binom{n_1}{u}\binom{n_2}{t-u},$$

where $u = r_1, t = r_1 + r_2$ and

$$f(u|t,\theta) = \frac{\binom{n_1}{u}\binom{n_2}{t-u}\theta^u}{\sum_j\binom{n_1}{j}\binom{n_2}{t-j}\theta^j}.$$

Using Theorem 5.2, the test for $H_0 : \theta = 1$ vs. $H_1 : \theta \neq 1$ is a similar test and is UMPU when randomization is used on the boundary.

Example 5.2

Consider $X \sim f(x) = \frac{e^{-\theta_1}\theta_1^x}{x!}$ and $Y \sim f(y) = \frac{e^{-\theta_2}\theta_2^y}{y!}$.

Test $H_0 : \theta_1 = \theta_2$ vs. $H_1 : \theta_1 \neq \theta_2$ or, equivalently, $H_0 : \frac{\theta_1}{\theta_2} = p = 1$ vs. $H_1 : p \neq 1$

$$f(x,y) = \frac{e^{-(\theta_1+\theta_2)}\theta_1^x\theta_2^y}{x!y!} = \frac{e^{-(\theta_1+\theta_2)}}{x!y!}e^{x\log\frac{\theta_1}{\theta_2}+(x+y)\log\theta_2}.$$

Let $\frac{\theta_1}{\theta_2} = p, x = u, x+y = t, \lambda = \theta_2$. Then

$$f(x,y) = \frac{e^{-(1+p)\lambda}}{u!(t-u)!}e^{u\log p+t\log\lambda},$$

which is in the form of Theorem 5.4. Hence a UMPU test exists based on

$$f(u|t) = \frac{e^{-(1+p)\lambda}}{u!(t-u)!} e^{u\log p + t\log\lambda}/f(t)$$

$$f(t) = \frac{e^{-(\theta_1+\theta_2)}(\theta_1+\theta_2)^t}{t!} = \frac{e^{-(1+p)\lambda}(1+p)^t\lambda^t}{t!}$$

$$f(u|t) = \binom{t}{u}\frac{p^u}{(1+p)^t} = \binom{t}{u}\left(\frac{p}{1+p}\right)^u\left(\frac{1}{1+p}\right)^{t-u},$$

which is binomial. For $\rho = \frac{p}{1+p}, \rho = \frac{1}{2}$ if $p = 1$. So the test is a UMPU level α test randomized on the boundary.

Example 5.3

Let $X_1,\ldots,X_n$ be i.i.d. from

$$f(x) = \frac{1}{\sigma}e^{-\frac{(x-\mu)}{\sigma}}$$

for $\mu > 0$ and $x \geq \mu$. Test

$$H_0\begin{cases} \mu = \mu_0 \\ \sigma = \sigma_0 \end{cases} \quad \text{vs.} \quad H_1\begin{cases} \mu = \mu_1 < \mu_0 \\ \sigma = \sigma_1 < \sigma_0 \end{cases}$$

We find the sufficient statistic and the UMP test. The likelihood is

$$L(\mu,\sigma) = \begin{cases} \sigma^{-n}e^{-\frac{n(\bar{x}-\mu)}{\sigma}} & x_{(1)} \geq \mu \\ 0 & \text{otherwise} \end{cases}$$

Note the sufficient statistic is $(x_{(1)}, \bar{x})$ and

$$\frac{L_0}{L_1} = \begin{cases} \left(\frac{\sigma_1}{\sigma_0}\right)^n e^{-\left[\frac{n(\bar{x}-\mu_0)}{\sigma_0} - \frac{n(\bar{x}-\mu_1)}{\sigma_1}\right]} & x_{(1)} \geq \mu_0 \\ 0 & x_{(1)} < \mu_0 \end{cases}$$

By the NP Lemma, reject H_0 if $L_0/L_1 \leq k_\alpha$. Therefore the UMP test rejects H_0 if

$$\bar{x} \leq \left\{\frac{1}{n}\log\left(k_\alpha\left(\frac{\sigma_0}{\sigma_1}\right)^n\right) + \left(\frac{\mu_1}{\sigma_1} - \frac{\mu_0}{\sigma_0}\right)\right\} \div \left(\frac{1}{\sigma_1} - \frac{1}{\sigma_0}\right) \equiv c_\alpha$$

or if $x_{(1)} < \mu_0$.

Example 5.4

Consider (X, Y) a bivariate normal vector with mean $(2\mu, 0)$ and covariance matrix

$$\Sigma = \begin{pmatrix} \mu^2 + 1 & -1 \\ -1 & 1 \end{pmatrix}, \quad \Sigma^{-1} = \frac{1}{\mu^2}\begin{pmatrix} 1 & 1 \\ 1 & 1+\mu^2 \end{pmatrix}, \quad |\Sigma| = \mu^2.$$

Here

$$f(x, y) = \frac{1}{2\pi\mu} e^{-\frac{1}{2\mu^2}\left[(x-2\mu,\ y)\begin{pmatrix} 1 & 1 \\ 1 & 1+\mu^2 \end{pmatrix}\begin{pmatrix} x-2\mu \\ y \end{pmatrix}\right]}$$

$$= \frac{1}{2\pi\mu} e^{-\frac{1}{2\mu^2}(x+y-2\mu)^2 - \frac{1}{2}y^2},$$

such that $t = \frac{X+Y}{2}$ is sufficient for μ and

$$t \sim N\left(\mu, \frac{\mu^2}{4}\right) \quad \text{and} \quad Y \sim N(0, 1).$$

Consider a one-sided test of $H_0 : \mu = \mu_0$ vs. $H_1 : \mu = \mu_1 > \mu_0$.

Now the MP test satisfies

$$\frac{L(\mu_0)}{L(\mu_1)} = \frac{\mu_1}{\mu_0} e^{-\frac{2(t-\mu_0)^2}{\mu_0^2} + \frac{2(t-\mu_1)^2}{\mu_1^2}} \le k_\alpha \text{ reject } \mu_0$$

$$\Longleftrightarrow \quad -\frac{2(t-\mu_0)^2}{\mu_0^2} + \frac{2(t-\mu_1)^2}{\mu_1^2} \le \log\left(k_\alpha \frac{\mu_0}{\mu_1}\right)$$

$$\Longleftrightarrow \quad t^2(\mu_0 + \mu_1) - 2t\mu_0\mu_1 \ge \frac{\mu_1^2\mu_0^2}{2(\mu_1 - \mu_0)}\log\left(k_\alpha \frac{\mu_1}{\mu_0}\right)$$

$$\Longleftrightarrow \quad (t - a_\alpha(\mu_1))(t - b_\alpha(\mu_1)) \ge 0, \quad a_\alpha > b_\alpha$$

$$\Longleftrightarrow \quad t \ge a_\alpha(\mu_1) \text{ or } t \le b_\alpha(\mu_1).$$

But this depends on μ_1 so there is no one-sided UMP test.
  Some remarks:

1. When a UMP test exists for scalar θ and some other conditions are satisfied then a scalar sufficient statistic exists.
2. UMP tests can exist even if a scalar sufficient statistic does not exist.
3. The existence of a scalar sufficient statistic does not even imply a one-sided UMP test.

Theorem 5.5 Suppose for $H_0 : \theta = \theta_0$ vs. $H_1 : \theta \neq \theta_0$

$$f(D|\theta) = C(\theta, \lambda)e^{\theta u(D) + \lambda t(D)} h(D)$$

and that $v = v(u, t)$ is independent of t for $\theta = \theta_0$. Then if $v(u, t) = a(t)u + b(t)$ with $a(t) > 0$ and increasing in u for each t, then

$$T(v) = \begin{cases} 1 & \text{when} & v < c_1 \text{ or } v > c_2 \\ p_i & \text{when} & v = c_i \ i = 1, 2 \\ 0 & \text{when} & c_1 < v < c_2 \end{cases}$$

is UMPU, where for $i = 1, 2$, c_i and p_i are determined by

(a) $E_{\theta_0}[T(v, t)|t] = \alpha$ and
(b) $E_{\theta_0}[T(v, t) \frac{(v - b(t))}{a(t)} |t] = \alpha E_{\theta_0}[\frac{v - b(t)}{a(t)} |t]$ or $E_{\theta_0}[vT(v, t)|t] = \alpha E_{\theta_0}[v|t]$ and since v
is independent of t for $\theta = \theta_0$ so are the c_i's and p_i's.

Example 5.5

For $X_1, \ldots, X_n$ i.i.d. $N(\mu, \sigma^2)$ and
$\qquad H_0 : \mu = \mu_0$ vs. $H_1 : \mu \neq \mu_0$ irrespective of σ^2,

$$L(\mu, \sigma^2) = (2\pi\sigma^2)^{-\frac{n}{2}} e^{-\frac{1}{2\sigma^2}[\Sigma(x_i - \mu)^2]}$$

$$= (2\pi\sigma^2)^{-\frac{n}{2}} e^{-\frac{n\mu^2}{2\sigma^2}} e^{-\frac{\Sigma x_i^2}{2\sigma^2} + \frac{\mu \Sigma x_i}{\sigma^2}}$$

$$= C(\theta, \lambda) \ e^{t\lambda + \theta u},$$

where

$$t = \Sigma x_i^2, \qquad \lambda = -\frac{1}{2\sigma^2}, \qquad \frac{n\mu}{\sigma^2} = \theta, \qquad u = \bar{x}.$$

So if we let $Y_i = X_i - \mu$, $E(Y) = \eta$ then the original hypothesis $H_0 : \mu = \mu_0$ becomes $H'_0 : \eta = 0$ and $H'_1 : \eta \neq 0$. Then we might as well start with $H'_0 : \mu = 0$.
$\qquad$ Let

$$V = \frac{\bar{X}}{\sqrt{\Sigma(X_i - \bar{X})^2}} = \frac{u}{\sqrt{t - nu^2}}.$$

Now for the normal distribution $N(\mu, \sigma^2)$, $\bar{X}$ and $s^2 = (n - 1)^{-1} \Sigma(X_i - \bar{X})^2$ are independent and in this case $\sqrt{n(n - 1)}V = \frac{\sqrt{n}\bar{X}}{s}$ is a student t with $n - 1$

degrees of freedom. The UMPU test for $H_0 : \mu = \mu_0$ vs. $H_1 : \mu \neq \mu_0$ is given as follows:

Let $t_{\alpha/2}$ be such that for t

$$\alpha/2 = \int_{-\infty}^{-t_{\alpha/2}} f(t_n)dt_n = \int_{t_{\alpha/2}}^{\infty} f(t_n)dt_n$$

and

$$t_{\alpha/2} = \sqrt{n(n-1)}v_{\alpha/2}.$$

Then reject H_0 if

$$v > \frac{1}{\sqrt{n(n-1)}}t_{\alpha/2} \quad \text{or} \quad v < -\frac{1}{\sqrt{n(n-1)}}t_{\alpha/2}.$$

Note also that the V or the student t is invariant under linear transformations. Let $aX_i + b = Z_i$, so that $\bar{Z} = a\bar{X} + b$ and

$$(N-1)s_z^2 = \Sigma(Z_i - \bar{Z})^2 = a^2\Sigma(X_i - \bar{X})^2 = a^2(N-1)s_x^2,$$

$$E\bar{Z} = aE(\bar{X}) + b = a\mu + b = \eta.$$

Therefore, $H_0 : \mu = \mu_0$ for x is equivalent to $H_0 : \eta = \eta_0$

$$\frac{\sqrt{n}(\bar{Z} - \eta_0)}{s_z} = \frac{\sqrt{n}(\bar{X} - \mu_0)}{s_x}.$$

Example 5.6

Suppose $X_1, \ldots, X_n$ are i.i.d. $N(\mu, \sigma^2)$ and $Y_1, \ldots, Y_m$ are i.i.d. $N(\eta, \sigma^2)$ and $H_0 : \mu = \eta$ vs. $H_1 : \mu \neq \eta$ with σ^2 unspecified. Then the UMPU test is based on student's t with $n + m - 2$ degrees of freedom where

$$t_{n+m-2} = \frac{\bar{X} - \bar{Y}}{s\sqrt{\frac{1}{n} + \frac{1}{m}}} \quad \text{for} \quad n\bar{X} = \sum_1^n X_i, \quad m\bar{y} = \sum_1^m Y_i \quad \text{and}$$

$$s^2 = \frac{\sum_1^n (X_i - \bar{X})^2 + \sum_1^m (Y_i - \bar{Y})^2}{n + m - 2}.$$

If $V_i = aX_i + b$ and $Z_i = aY_i + b$ then for H_0 vs. H_1

$$E(V) = a\mu + b, \quad E(Z) = a\eta + b, \quad a \neq 0, \quad \text{var}(V) = a^2\sigma^2 = \text{var}(Z).$$

Hence

$$\frac{\bar{V} - \bar{Z}}{\sqrt{\left(\frac{1}{n} + \frac{1}{m}\right)(\Sigma(V_i - \bar{V})^2 + \Sigma(Z_i - \bar{Z})^2)/(n + m - 2)}}$$

$$= \frac{a(\bar{X} - \bar{Y})}{\sqrt{a^2 s^2 \left(\frac{1}{n} + \frac{1}{m}\right)}} = t_{n+m-2}.$$

Since the UMPU test requires $\alpha = \frac{\alpha}{2} + \frac{\alpha}{2}$ from both tails of the t-test, the test is also invariant as to whether we consider t_{n+m-2} or $-t_{n+m-2}$ that is, $|t|$ or even t^2. Note also it leaves the problem invariant in that

$$H_0 : \mu = \eta \text{ is equivalent to } H_0' : a\mu + b = a\eta + b$$
$$H_1 : \mu \neq \eta \text{ is equivalent to } H_1' : a\mu + b \neq a\eta + b.$$

5.4 INVARIANT TESTS

More generally suppose we have k normal populations with assumed $N(\mu_i, \sigma^2)$ $i = 1, \ldots, k$ and random samples of size n_i from each of the populations. Recall for testing $H_0 : \mu_1 = \mu_2 = \cdots = \mu_k$ vs. H_1 : not all the μ_i's are equal and σ^2 unspecified, that the F-test used to test this was

$$F_{k-1, n-k} = \frac{\Sigma n_i(\bar{X}_i - \bar{X})^2/(k-1)}{s^2}, \quad n_i \bar{x}_i = \sum_{j=1}^{n_i} X_{ij}, \quad n\bar{x} = \sum_{1}^{k} n_i \bar{X}_i,$$

$$n = \sum_{1}^{k} n_i$$

and

$$(n-k)s^2 = \sum_{i=1}^{k} \sum_{j=1}^{n_i} (X_{ij} - \bar{X}_i)^2.$$

If $Y_{ij} = aX_{ij} + b$ for $a \neq 0$, then

$$\frac{\Sigma n_i(\bar{Y}_i - \bar{Y})^2/(k-1)}{s_y^2} = F_{k-1, n-k},$$

as previously defined. The statistic is invariant under linear transformations. Note also that $H_0 : \mu_1 = \cdots = \mu_k$ is equivalent to $H_0' : a\mu_i + b = \cdots = a\mu_k + b$ and $H_1 : \implies$ to H_1'. So it is an invariant test and it turns out to be UMP among invariant tests so we say it is UMPI. However this test is not UMPU.

In general, for the problem of

$$H : \theta \in \Theta_H \quad \text{vs.} \quad K : \theta \in \Theta_K$$

for any transformation g on D which leaves the problem invariant, it is natural to restrict our attention to all tests $T(D)$ such that $T(gD) = T(D)$ for all $D \in S$. A transformation g is one which essentially changes the coordinates and a test is invariant if it is independent of the particular coordinate system in which the data are expressed.

We define invariance more precisely:

Definition: If for each $D \in S$ the function $t(D) = t(gD)$ for all $g \in G$, G a group of transformations then t is invariant with respect to G.

Recall that a set of elements G is a group if under some operation it is

(a) *closed*: for all g_1 and $g_2 \in G$, $g_1 g_2 \in G$;

(b) *associative*: $(g_1 g_2)g_3 = g_1(g_2 g_3)$ for all g_1, g_2, $g_3 \in G$;

(c) *has an identity element*: $g_I g = g g_I = g$ where $g_I \in G$;

(d) *has an inverse*: that is, if $g \in G$ then $g^{-1} \in G$ where $g g^{-1} = g^{-1} g = g_I$.

Theorem 5.6   Let $y = gD$ and for each point D the set of values y as g runs through all the elements of G is called the orbit traced out by D. Then it is necessary and sufficient that t be constant on each of its orbits for t to be invariant.

Proof: First assume t is invariant that is, $t(gD) = t(D)$ for each $D \in S$ and all $g \in G$. Therefore, t is constant on all of its orbits. Conversely if t is constant on all of its orbits, then $t(gD) =$ constant for each D and all g. Since it is true for all g then it is also true for g_I where $g_I D = D$ and constant $= t(gD) = t(g_I D) = t(D)$ as required.   □

Theorem 5.7   A function $t(D)$ is defined as a maximal invariant if it is invariant (constant on all its orbits $t(D) = t(gD)$) and for each orbit takes on a different value or if $t(D) = t(D') \Rightarrow D' = gD$ for some $g \in G$. Further if $t(D)$ is a maximal invariant with respect to G then any test that is a function of t, $T(D) = f(t(D))$ is invariant for all D and conversely if $T(D)$ is an invariant test then it depends on the maximal invariant $t(D)$.

Proof: Let $T(D) = f(t(D))$, $t(D)$ is the maximal invariant, then $T(gD) = f(t(gD)) = f(t(D)) = T(D)$ such that T is invariant. Conversely if T is invariant and $t(D)$ is a maximal invariant then

$$T(D) = T(gD) \text{ and } t(D) = t(D') \text{ implies } D' = gD$$

for some g then $T(D) = T(D')$ so that T depends on t.

Hence for a class of invariant tests we need only consider the maximal invariant.

Let $D \sim F_\theta$ for $\theta \in \Theta$ and let g be a one-to-one transformation of the sample space S onto itself. Let gD be the random variable that transforms D by g and hence has distribution $F_{\theta'}$ and assume $\theta' \in \Theta$ and all $\theta' = g'\theta$ is the induced transformation. Now we say that the parameter set Θ remains invariant (is preserved) under g if $g'\theta \in \Theta$ for all $\theta \in \Theta$ and if in addition for any $\theta' \in \Theta$ there exists a $\theta \in \Theta$ such that $g'\theta = \theta'$. We express this by

$$g'\Theta = \Theta.$$

The transformation of Θ onto itself then is one-to-one provided that F_θ corresponding to different values of θ are distinct. Further we say that the problem of testing $H : \theta \in \Theta_H$ vs. $K : \theta \in \Theta_K$ remains invariant under a transformation g if g' preserves both Θ_H and Θ_K that is,

$$g'\Theta_H = \Theta_H \quad \text{and} \quad g'\Theta_K = \Theta_K.$$

If $\mathcal{G}$ is a class of transformations that leave the problem invariant and G the smallest class of transformations containing $\mathcal{G}$ that is a group, then G will also leave the problem invariant. □

Example 5.7

Suppose $X_i, \ldots, X_n$ are i.i.d. $N(\mu, \sigma^2)$ and $Y_1, \ldots, Y_n$ are i.i.d. $N(\mu, \tau^2)$. Consider

$$H_0 : \sigma^2 = \tau^2 \quad \text{or} \quad \frac{\sigma^2}{\tau^2} = 1 = \frac{\tau^2}{\sigma^2},$$

$$H_1 : \sigma^2 \neq \tau^2 \qquad \frac{\sigma^2}{\tau^2} \neq 1 \neq \frac{\tau^2}{\sigma^2}.$$

The transformation

$$X_i' = aX_i + b \quad \text{and} \quad Y_i' = aY_i + c, \quad a \neq 0, \ X_i' = Y_i, \ Y_i' = X_i,$$

is such that $\text{var} X_i' = a^2 \text{var} X_i$, $\text{var} Y_i' = a^2 \text{var} Y_i$ so

$$H_0' : \frac{a^2 \sigma^2}{a^2 \tau^2} = 1, \quad H_1' : \frac{a^2 \sigma^2}{a^2 \tau^2} \neq 1.$$

Under the transformations $X_i' = Y_i$ and $Y_i' = X_i$,

$$H_0'' : \frac{a^2 \tau^2}{a^2 \sigma^2} = 1 \ \text{iff} \ \frac{\tau^2}{\sigma^2} = 1 \quad H_1'' : \frac{a^2 \tau^2}{a^2 \sigma^2} \neq 1 \ \text{iff} \ \frac{\tau^2}{\sigma^2} \neq 1.$$

So the problem is invariant under the transformations, that is, a linear transformation and a permutation. Consider

$$T(D) = 1 \quad \text{if } t = \text{Max}[Z, \, Z^{-1}] > k_\alpha$$
$$\qquad\quad = 0 \quad \text{otherwise}$$

$$\text{for } Z = \frac{\Sigma(X_i - \bar{X})^2}{\Sigma(Y_i - \bar{Y})^2}.$$

Note that

$$Z' = \frac{a^2\Sigma(X_i - \bar{X})^2}{a^2\Sigma(Y_i - \bar{Y})^2} = Z,$$

$$\frac{\Sigma(Y'_i - \bar{Y}')^2}{\Sigma(X'_i - \bar{X}')^2} = \frac{\Sigma(X_i - \bar{X})^2}{\Sigma(Y_i - \bar{Y})^2} = Z,$$

and

$$\frac{\Sigma(X'_i - \bar{X}')^2}{\Sigma(Y'_i - \bar{Y}')^2} = \frac{\Sigma(Y_i - \bar{Y})^2}{\Sigma(X_i - \bar{X})^2} = Z^{-1}.$$

Hence t is invariant under the transformations and $T(D)$ is UMP among invariant tests. It can be made UMPU. Usually we use equal tail probabilities which bias the test. For unequal sample sizes and a one sided alternative the test is also UMPU and UMPI.

Example 5.8

Consider

$$X_{ij} \sim N(\mu_i, \sigma_i^2) \quad i = 1, \ldots, k, \quad j = 1, \ldots, n_i,$$

and let

$$H_0 : \sigma_1^2 = \cdots = \sigma_k^2 \text{ vs. } H_1 : \text{one or more of the } \sigma_i^2 \text{ are different.}$$

For $k > 2$ no UMPU or UMPI test exists. A UMPI test may actually be inadmissible as the following example by Stein shows.

Example 5.9

Let

$$X_1 = \begin{pmatrix} X_{11} \\ X_{12} \end{pmatrix}, \quad X_2 = \begin{pmatrix} X_{21} \\ X_{22} \end{pmatrix}, \quad X_1 \sim N_2\left(\begin{pmatrix} 0 \\ 0 \end{pmatrix}, \Sigma\right), \quad X_2 \sim N_2\left(\begin{pmatrix} 0 \\ 0 \end{pmatrix}, \Delta\Sigma\right)$$

and

$$\Sigma = \begin{pmatrix} \sigma_1^2 & \rho\sigma_1\sigma_2 \\ \rho\sigma_1\sigma_2 & \sigma_2^2 \end{pmatrix}$$

is non-singular. To test

$$H : \Delta = 1 \quad \sigma_1^2, \sigma_2^2, \rho \text{ are unspecified, versus}$$
$$K : \Delta > 1 \quad \sigma_1^2, \sigma_2^2, \rho \text{ are unspecified,}$$

we note that for any non-singular transformation A

$$Y_1 = AX_1, \quad Y_2 = AX_2 \quad EY_1 = 0 = EY_2,$$

$Cov(Y_1) = A\Sigma A'$, $Cov(Y_2) = \Delta A\Sigma A'$ that H and K are invariant under all A. We now inquire as to the totality of invariant tests T under a non-singular transformation. Now the set of data points:

$$D = (X_1, X_2) = \begin{pmatrix} x_{11} & x_{21} \\ x_{12} & x_{22} \end{pmatrix}$$

is non-singular with probability 1 since $P(X_{11}X_{22} = X_{21}X_{22}) = 0$. Therefore the sample space is the set of all such real non-singular matrices D. Further for any 2 such matrices D and D' there exists a non-singular transformation A such that $D' = AD$. Since for any $D \in S$ as we go through all non-singular transformations A, the orbit traced out is the whole sample space S. Hence there is a single orbit and the only invariant transformation $t(D)$ has to be constant over the single orbit or since $T(D) = f(t(D))$ this implies $T(D) = $ constant for all D. Hence the only invariant test of size α is $T(D) \equiv \alpha$. It then follows vacuously that among all invariant tests it is UMP or UMPI and consequently useless.

Now X_{11} and X_{21} are independently distributed as $N(0, \sigma_1^2)$ and $N(0, \Delta\sigma_1^2)$. Hence

$$\frac{X_{11}^2}{\sigma_1^2} \sim \chi_1^2 \qquad \frac{X_{21}^2}{\Delta\sigma_1^2} \sim \chi_1^2$$

are independent chi-square variables with one degree of freedom. Let $U = X_{21}^2/X_{11}^2$ and reject H if $U \le k_\alpha$. Then

$$\frac{X_{21}^2/\sigma_1^2\Delta}{X_{11}^2/\sigma_1^2} \sim F(1, 1),$$

where $F(1, 1)$ is the F variate with degrees of freedom $(1, 1)$ and $U \sim \Delta F(1, 1)$ under K. Now

$$\alpha = P(U \le F_\alpha(1, 1) | \Delta = 1)$$

but

$$1 - \beta(\Delta) = P[U \leq \Delta F_\alpha(1, 1)|\Delta > 1) > \alpha$$

and $1 - \beta(\Delta)$ increases monotonically as Δ increases. Hence the UMPI test is inadmissible in that it is dominated by this test.

Example 5.10

A UMPI level $\alpha = 0.1$ test can be obtained for H vs. K, Hacking (1965), where probabilities for H and K obtain as in Table 5.1.

First note that at level 0.1

$$\frac{L(H|D = 0)}{L(K_j|D = 0)} = \frac{0.90}{0.91},$$

while

$$\frac{L(H|D = j)}{L(K_j|D = j)} = \frac{1}{90}, \quad j = 1, \ldots, 100,$$

$$\frac{L(H|D \neq j, 0)}{L(K_j|D \neq j, 0)} = \frac{0.001}{0} = \infty.$$

Table 5.1: Probabilities for H and K

	D	0	1	2	3	·	·	·	·	j	·	·	·	99	100
	H	.90	.001	.001	.001	·	·	·	·	.001	·	·	·	.001	.001
	K_1	.91	.09	0	0	·	·	·	·	0	·	·	·	0	0
	K_2	.91	0	.09	0	·	·	·	·	0	·	·	·	0	0
	K_3	.91	0	0	.09	0	·	·	·	0	·	·	·	0	0
	·	·	·	·	0	·	·	·	·	·	·	·	·	·	·
	·	·	·	·	·	·	·	·	·	·	·	·	·	·	·
K	·	·	·	·	·	·	·	·	·	·	·	·	·	·	·
	·	·	·	·	·	·	·	·	·	0	·	·	·	·	·
	K_j	.91	0	0	0	·	·	·	0	.09	0	·	·	0	0
	·	·	·	·	·	·	·	·	·	0	·	·	·	·	·
	·	·	·	·	·	·	·	·	·	·	·	·	·	·	·
	·	·	·	·	·	·	·	·	·	·	·	·	·	·	·
	·	·	·	·	·	·	·	·	·	·	·	·	·	·	·
	K_{100}	.91	0	0	0	·	·	·	·	0	·	·	·	0	.09

A randomized MP test of H vs. K_j at level $\alpha = 0.1$ is

$$T_j(D) = \begin{cases} 1 & \text{if } D = j \quad j = 1, \ldots, 100 \\ 0.11 & \text{if } D = 0 \\ 0 & \text{if } D \neq 0, j \end{cases}$$

since

$$\alpha = 0.001 + 0.11 \times 0.90 = .1$$
$$1 - \beta = 0.09 + 0.11 \times 0.91 = 0.1901.$$

However, no UMP test for H vs. K exists since it depends on K_j.
If we used the test for K_j on K_t

$$1 - \beta_t = 0 + 0.91 \times 0.11 = 0.1001,$$

which has less power than using it for K_j. Intuitively it would appear if $D = 0$ is observed there is very little to discriminate between H and K, but if $D \neq 0$ occurred, say $D = 2$, then while we can rule out all K_j for $j \neq 2, K_2$ has 90 times the support of H. Therefore, at first glance, it would appear reasonable to reject H if $D > 0$ occurred. Consider the test

$$T(D) = \begin{cases} 1 & \text{if } D = j \quad \text{for } j = 1, \ldots, 100 \\ 0 & \text{if } D = 0, \end{cases}$$

such that $\alpha = 0.1$ and $1 - \beta = 0.09$ so this "reasonable" test is "worse than useless" since for $\alpha = 0.1$ more power can be obtained for $T(D) \equiv 0.1$. This is the likelihood test, but the N-P UMPI test is

$$T^*(D) = \begin{cases} \dfrac{1}{9} & D = 0 \\ 0 & D = j \quad \text{for } j = 1, \ldots, 100, \end{cases}$$

where $\alpha = \frac{1}{9} \times 0.9 = 0.1, 1 - \beta = \frac{1}{9} \times 0.91 = 0.101$.
To show that this is the UMPI test we note, due to the symmetry of the problem, that it is clear that any transformation of H and K must send $0 \to 0$ and $(1, \ldots, 100) \to (i_1, \ldots, i_{100})$ where $(i_1, \ldots, i_{100})$ is a permutation of $(1, \ldots, 100)$. Hence invariant tests must treat $D = 0$ in one way and $D = 1, \ldots, 100$ in the same way so that

$$\alpha = 0.1 \geq E_H(T(D)) = T(0)P(D = 0|H) + T(D \neq 0)P(D \neq 0|H)$$
$$= T(0) \times 0.9 + T(D \neq 0) \times 0.1 = 0.9p + 0.1q$$
$$1 - \beta = E_K(T(D)) = T(0)P(D = 0|K) + T(D \neq 0)P(D \neq 0|K)$$
$$= 0.91p + 0.09q = \max.$$

We can add to p and q until $.9p + 0.1q = 0.1$ and then solve for q from $9p + q = 1$. Hence

$$q = 1 - 9p, \quad 0 \le p \le \frac{1}{9}$$

$$0.91p + 0.09(1 - 9p) = \max$$

$$0.1p + 0.09 = \max.$$

The left-hand side is an increasing function of p then we set $p = \frac{1}{9}$ and hence $q = 0$ therefore $1 - \beta = 0.91 \times \frac{1}{9} = 0.1011$ and

$$T(j) = \begin{cases} \dfrac{1}{9} & j = 0 \\ 0 & j \ne 0 \end{cases}$$

is the UMPI test.

Now notice that the likelihood test

$$T(D) = \begin{cases} 0 & \text{if} \quad D = 0 \\ 1 & \text{if} \quad D = 1, \ldots, 100 \end{cases}$$

is also invariant and has minimum power among all level $\alpha = .1$ tests, since for $9p + q = 1$ and $.1p + .09 = $ minimum for $p = 0$ and has power $.09$ so the likelihood test is a uniformly least powerful invariant test (ULPI). Note also that the tests UMPI and ULPI never agree to reject H, so they are contradictory. Also if we are using the UMPI test for evaluating H in the light of the result of a trial it would be very strange since if say $D = 2$ occurred we would know that only K_2 among all the K_j's was possible so that a MP test would reject H for K_2 but the UMPI test tells us to reject K. This is not a paradox since the N-P test theory is based on a pre-trial assessment and before the trial the UMPI test is best.

Remark: If no UMPU or UMPI test exists the next step is to look for a UMP among all tests that are both unbiased and invariant and this would be a UMPUI test.

5.5   LOCALLY BEST TESTS

When there are no UMP tests we may sometimes restrict the alternative parameter values to cases of presumably critical interest and look for high power against these alternatives. In particular if interest is focused on alternatives close to H_0: say $\theta = \theta_0$, we could define $\delta(\theta)$ as a measure of the discrepancy of a close alternative from H_0.

Definition: A level α test T is defined as locally most powerful (LMP) if for every other test T^* there exists a Δ such that

$$1 - \beta_T(\theta) \ge 1 - \beta_{T^*}(\theta) \quad \text{for all } \theta \text{ such that } 0 < \delta(\theta) < \Delta.$$

Theorem 5.8   If among all unbiased level α tests T^*, T is LMP then we say T is LMPU.

1. For real θ for any T^* such that $1 - \beta_{T^*}(\theta)$ is continuously differentiable at $\theta = \theta_0$ with $H_1 : \theta > \theta_0$ or $H_1 : \theta < \theta_0$ (i.e., one sided tests) a LMP test exists and is defined such that for all level α tests T^* there is a unique T such that

$$\arg\max_{T^*} \left[\frac{d(1 - \beta_{T^*}(\theta))}{d\theta} \right]_{\theta=\theta_0} = T.$$

2. For θ real valued and $1 - \beta_{T^*}(\theta)$ twice continuously differentiable at $\theta = \theta_0$ for all T^*, then a LMPU level α test T exists of $H_0 : \theta = \theta_0$ vs. $H_1 : \theta \neq \theta_0$ and is given by

$$\arg\max_{T^*} \left[\frac{d^2(1 - \beta_{T^*}(\theta))}{d\theta^2} \right]_{\theta=\theta_0} = T \implies 1 - \beta_T(\theta) \geq 1 - \beta_{T^*}(\theta)$$

for $0 < \delta(\theta) < \Delta$. All locally unbiased tests result in

$$\frac{d[1 - \beta_{T^*}(\theta)]}{d\theta}\bigg|_{\theta=\theta_0} = 0$$

given the condition of being continuously differentiable.

Proof of (1): For a test T^*

$$\gamma_{T^*}(\theta) \equiv 1 - \beta_{T^*}(\theta) = \int T^* f(D|\theta) d\mu,$$

$$\gamma'_{T^*}(\theta) = \int T^* \frac{\partial f_\theta}{\partial \theta} d\mu.$$

Suppose $\gamma'_T(\theta_0) \geq \gamma_{T^*}(\theta_0)$. Then note that

$$\gamma_T(\theta_0) = 1 - \beta_T(\theta_0) = 1 - \beta_{T^*}(\theta_0) = \gamma_{T^*}(\theta_0)$$

$$\gamma'_T(\theta_0) = \lim_{\Delta\theta \to 0} \frac{\gamma_T(\theta_0 + \Delta\theta) - \gamma_T(\theta_0)}{\Delta\theta} \geq \lim_{\Delta\theta \to 0} \frac{\gamma_{T^*}(\theta_0 + \Delta\theta) - \gamma_{T^*}(\theta_0)}{\Delta\theta} = \gamma'_{T^*}(\theta_0)$$

$$\gamma'_T(\theta_0) - \gamma'_{T^*}(\theta_0) = \lim_{\Delta\theta \to 0} \frac{\gamma_T(\theta_0 + \Delta\theta) - \gamma_{T^*}(\theta_0 + \Delta\theta)}{\Delta\theta} \geq 0$$

such that for some $\Delta\theta > 0$

$$\gamma_T(\theta_0 + \Delta\theta) \geq \gamma_{T^*}(\theta_0 + \Delta\theta)$$

or

$$1 - \beta_T(\theta) \geq 1 - \beta_{T^*}(\theta) \quad \text{for } \theta_0 \leq \theta \leq \theta_0 + \Delta\theta.$$

Then

$$\alpha = \gamma_{T^*}(\theta_0) = \int T^* f_{\theta_0} d\mu = \int T f_{\theta_0} d\mu = \gamma_T(\theta_0)$$

and

$$\int (T - T^*) f_{\theta_0} d\mu = 0.$$

For $k > 0$

$$0 \leq k[\gamma_T'(\theta_0) - \gamma_{T^*}'(\theta_0)] = k \int (T - T^*) f_{\theta_0}' d\mu$$

$$- \int (T - T^*) f_{\theta_0} d\mu$$

$$= \int (T - T^*) [k f_{\theta_0}' - f_{\theta_0}] d\mu.$$

Now if

$$T(D) = \begin{cases} 1 & \text{for} & f_{\theta_0} < k f_{\theta_0}' \\ p & \text{for} & f_{\theta_0} = k f_{\theta_0}' \\ 0 & \text{for} & f_{\theta_0} > k f_{\theta_0}' \end{cases},$$

and defining $s_1 \cup s_2 = S$ in the table below

	$f_{\theta_0} < k f_{\theta_0}'$	$f_{\theta_0} > k f_{\theta_0}'$
$T > T^*$	s_1	$\emptyset$
$T < T^*$	$\emptyset$	s_2

then

$$0 \leq \int_S (T - T^*)(k f_{\theta_0}' - f_{\theta_0}) d\mu = k(\gamma_T'(\theta_0) - \gamma_{T^*}'(\theta_0))$$

and the N-P lemma is satisfied. If the inequality holds then T is unique and has maximum power in an interval around θ_0 and is LMP.

For a Locally Most Powerful Unbiased (LMPU) test Neyman and Pearson (1936–1938) show that

$$
T(D) = \begin{cases} 1 & \text{if} \\ p & \text{if} \\ 0 & \text{if} \end{cases} \quad \begin{matrix} k_1 f(D|\theta_0) + k_2 f'(D|\theta_0) & < & f''(D|\theta_0) \\ & = & \\ & > & \end{matrix}
$$

where k_1 and k_2 are determined to satisfy

$$
E[T(D)|\theta_0] = \alpha, \qquad \int f'(D|\theta_0)d\mu = 0,
$$

is LMPU.

5.6 TEST CONSTRUCTION

So far N-P theory has not really given a principle for constructing a test. It has indicated how we should compare tests that is, for a given size the test with the larger power is superior, or for sample space S, we want $T(D)$ to be such that for all tests $T^*(D)$

$$
E(T^*|H) \le \alpha
$$

choose $T(D)$ such that

$$
E(T^*|K) \le E(T|K)
$$

and as this doesn't always happen we go on to other criteria, unbiasedness, invariance, and so forth.

There are several test construction methods that are not dependent on the N-P approach but are often evaluated by the properties inherent in that approach. The most popular one is the Likelihood Ratio Test (LRT) criterion.

Specifically the Likelihood Ratio Test (LRT) criterion statistic for a set of parameters θ to test H_θ vs. K_θ is defined as

$$
\frac{\sup_{\theta \in H_\theta} L(\theta|D)}{\sup_{\theta \in (H_\theta \cup K_\theta)} L(\theta|D)} = \lambda(D) \le 1
$$

with critical region defined as $\lambda < k_\alpha$ reject H_θ. This yields

$$
P\{\lambda(D) < k_\alpha\} \le \alpha.
$$

A variation is

$$\frac{\sup_{\theta \in H_\theta} L(\theta_1 | D)}{\sup_{\theta \in K_\theta} L(\theta | D)} = \lambda^*(D),$$

so that if $\lambda^* < 1$ then $\lambda = \lambda^*$, if $\lambda^* \geq 1$ then $\lambda = 1$.

So whenever the rejection region is $\lambda < k_\alpha \leq 1$ they are identical. Now for a simple null versus a simple alternative it is essentially a likelihood test and by N-P it is MP or UMP. More generally if a UMP test exists it is expected that the LRT will be UMP, but this remains to be shown. At any rate UMP tests are rare.

If a UMP does not exist and we look for a UMPU test the LRT will on occasion produce a biased test but sometimes a simple bias adjustment that is, substitute unbiased estimates for the maximum likelihood estimates, will correct this. Even UMPU tests are rare and in the general linear hypothesis we often get a UMPI test by virtue of the LRT criterion and the LRT would seem to be valuable where UMP and UMPU tests do not exist. Although a LRT is often biased it has a property, under rather general conditions for example, $d^2 \log f(x|\theta)/d\theta^2$ exists and is dominated by an integrable function, that it is consistent in the sense that for $D_n(x_1, \ldots, x_n)$ and $P(\lambda(D_n) < \lambda_{\alpha,n} | H_\theta) \leq \alpha$ then

$$\lim_{n \to \infty} P(\lambda(D_n) < \lambda_{\alpha, n} | K_\theta) = \lim_{n \to x} (1 - \beta_{n, K_\theta}) = 1$$

that is, for any member of the alternative K_θ the power tends to 1 as n increases or the probability of rejecting a false hypothesis increases with n and tends to certainty. This also shows that the LRT is asymptotically unbiased although an unbiased test need not be consistent.

An LRT also has the property that under "pleasant regularity conditions"

$$-2 \log \lambda(D_n) \longrightarrow \chi_r^2$$

a chi-square with r degrees of freedom under H_0 where

$$\theta = (\theta_1, \ldots, \theta_r, \theta_{r+1}, \ldots, \theta_s) = (\theta^{(r)}, \theta^{(s-r)})$$
$$\theta_0^{(r)} = (\theta_{10}, \ldots, \theta_{r0})$$
$$H_0 : \theta^{(r)} = \theta_0^{(r)}, \quad \theta^{(n-r)} \text{ unspecified}$$
$$H_1 : \theta^{(r)} \neq \theta_0^r, \quad \theta^{(n-r)} \text{ unspecified.}$$

Example 5.11

Let $X_1, X_2, \ldots, X_n$ i.i.d. with

$$f(x|\theta) = \frac{1}{\pi(1 + (x - \theta)^2)}.$$

Consider

$$H_0 : \theta = 0, \quad \text{versus} \quad H_1 : \theta \neq 0.$$

It is easy to show that a test based on $\bar{X}$ is unbiased but not consistent.

An example of a worse than useless LRT, reported by Lehmann (1950), is demonstrated in the following:

Example 5.12

Consider the following table of probabilities under H_0 and H_1:

D	-2	-1	0	1	2
$P(D|H_0)$	$\alpha/2$	$\frac{1}{2} - \alpha$	α	$\frac{1}{2} - \alpha$	$\alpha/2$
$P(D|H_1)$	$\theta_1(1 - \theta_2)$	$\left(\frac{1}{2} - \alpha\right)\left(\frac{1-\theta_1}{1-\alpha}\right)$	$\alpha\left(\frac{1-\theta_1}{1-\alpha}\right)$	$\left(\frac{1}{2} - \alpha\right)\left(\frac{1-\theta_1}{1-\alpha}\right)$	$\theta_1\theta_2$

and the following null and alternative hypotheses

$$H_0 : \alpha < \frac{1}{2} \text{ and known } \theta_1 = \alpha, \quad \theta_2 = \frac{1}{2}$$

$$H_1 : 0 \leq \theta_1 < \alpha < \frac{1}{2}, \quad 0 \leq \theta_2 \leq 1, \quad \theta_2 \neq \frac{1}{2}.$$

The LRT of level α is based on

$$\lambda(D) = \frac{L(H_0|D)}{\sup L(H_1|D)}$$

D	-2	-1	0	1	2
$L(H_0|D)$	$\alpha/2$	$\frac{1}{2} - \alpha$	α	$\frac{1}{2} - \alpha$	$\alpha/2$
$\sup L(H_1|D)$	α	$\left(\frac{1}{2} - \alpha\right)/(1 - \alpha)$	$\frac{\alpha}{1-\alpha}$	$\left(\frac{1}{2} - \alpha\right)/(1 - \alpha)$	α
LRT $\lambda(D)$	$1/2$	$1 - \alpha$	$1 - \alpha$	$1 - \alpha$	$1/2$

First note the following level α test which is not the LRT. Define T such that

$$T(D) = \begin{cases} 1 & \text{when } D = 0 \\ 0 & D = \pm 1 \text{ and } \pm 2. \end{cases}$$

Since $\alpha < \frac{1}{2}$ then $1 - \alpha > \frac{1}{2}$ and T has size α and power $\alpha(\frac{1-\theta_1}{1-\alpha}) > \alpha$ since $1 - \theta_1 > 1 - \alpha$.

Now consider the LRT:

$$T(D) = \begin{cases} 1 & \text{if} \quad \lambda(D) < 1 - \alpha \quad \text{or} \quad D = -2, 2 \\ 0 & \text{if} \quad \lambda(D) \geq 1 - \alpha \quad \text{or} \quad D = -1, 0, 1. \end{cases}$$

This results in level $= P(D = 2|H_0) + P(D = -2|H_0) = \frac{\alpha}{2} + \frac{\alpha}{2} = \alpha$ and power $= P(D = 2|H_1) + P(D = -2|H_1) = \theta_1(1 - \theta_2) + \theta_1\theta_2 = \theta_1 < \alpha$ such that the LRT is "worse than useless." Notice also that LRT is also a likelihood test.

Perhaps we are paying too much attention to size and power which are pre-test evaluations and not necessarily appropriate for post-trial evaluations.

5.7 REMARKS ON N-P THEORY

A synopsis of criticisms of N-P theory due to Hacking (1965) follows. The N-P hypothesis testing theory is then one of fixing a small size and searching for large power. The rationale behind it is to search for rules for governing our behavior with regard to hypotheses (without hoping to know whether any one of them is true or false) which will ensure that in the long run we shall not be wrong too often. To assert whether H be rejected or not, calculate D (the observables) and if $D \in s$ reject H, if $D \notin s$ accept H. Such a rule tells us nothing as to whether in a particular case H is true when $D \notin s$ or false when $D \in s$. If we behave in such a way we shall reject when it is true not more than 100 $\alpha\%$ of the time and in addition we may have evidence that we shall reject H sufficiently often when it is false.

Presumably if we behave in such a way and keep α fixed we shall reject hypotheses tested through our lifetimes that are true not more than 100 $\alpha\%$ of the time (i.e., there is a very high likelihood that this will happen in the long run), and really one is no more or less certain about any of these hypotheses. But if we had to adopt a testing policy now and were bound to follow it for the rest of our lives so that for every false hypothesis we rejected we would have bestowed upon us h heavenly units and likewise for every true hypothesis we accept, while we lose the same for each true one we reject and every false one we accept then this is the best life long policy—but no one has ever been in this situation. This is of course a pre-trial, not a post trial evaluation.

Now it may be that before a trial a decision must be made (because accuracy is difficult or tedious or it may be economical to do so) only to note whether a trial made is in s or not. This may be wholly rational and in this case N-P theory is an economical post trial evaluation. Of course in some cases, say, a simple H_0 vs. a simple H_1, where a MP test exists, the extra knowledge will not change our evaluations of H_0. In these cases where a rational decision to discard or ignore data the N-P theory is a special case of likelihood.

In any event the N-P theory can be viewed from a likelihood perspective:

$$P(D \in s|H) = L(H|D \in s) \leq \alpha(H) \quad \text{say small relative to}$$
$$P(D \in s|K) = L(K|D \in s) = 1 - \beta(K) \quad \text{so}$$

$$Q = \frac{L(H|D \in s)}{L(K|D \in s)} \leq \frac{\alpha(H)}{1 - \beta(K)} \quad \text{small.}$$

Another way of looking at N-P theory is that the N-P testing program (small size, large power) is well supported before a trial is made and data observed. Consider the following metahypotheses:

M_1: For the kind of experiment to be made the test will misclassify H that is, reject H when true or accept H when false.

M_2: For the kind of experiment to be made the test will correctly classify H that is, accept when true and reject when false.

In view of the low size and large power of the test M_2 is much better supported than M_1 but only before a trial is made. Since we are dealing with the long run let p be the fraction of the time H is true. Then for $\bar{H}$ representing H false,

$$P(\text{reject } H|H) = \alpha, \qquad P(\text{accept } H|H) = 1 - \alpha$$
$$P(\text{reject } H|\bar{H}) = 1 - \beta, \qquad P(\text{accept } H|\bar{H}) = \beta$$
$$R = \frac{P(M_2|I)}{P(M_1|I)} = \frac{p(1-\alpha) + (1-p)(1-\beta)}{p\alpha + (1-p)\beta}.$$

If $\alpha < \beta$ then $1 - \alpha > 1 - \beta$ and

$$R > \frac{p(1-\beta) + (1-p)(1-\beta)}{p\beta + (1-p)\beta} = \frac{1-\beta}{\beta}.$$

If $\alpha > \beta$ then

$$R > \frac{p(1-\alpha) + (1-p)(1-\alpha)}{p\alpha + (1-p)\alpha} = \frac{1-\alpha}{\alpha}.$$

So M_2 is well supported relative to M_1, assuming low size and large power.

But if we actually observe $D \in s$ so that H is less well supported than K we should not necessarily reject H because the actual value of D may be an observation which cannot occur if H is false.

5.8 FURTHER REMARKS ON N-P THEORY

We now reconsider an example presented earlier.

Example 3.2 (*continued*)

Suppose we make independent trials of Binary variables with probability p then if we hold n = number trials fixed and got r successes then

$$P(R = r|n) = \binom{n}{r} p^r (1-p)^{n-r}.$$

For $n = 5$ $H_0 : p = \frac{1}{2}$ $H_1 : p = p_1 < \frac{1}{2}$ and $\alpha = \frac{1}{16}$ a UMP N-P test is

$$T(r) = \begin{cases} 1 & \text{if} \quad r = 0 \\ \dfrac{1}{5} & \text{if} \quad r = 1 \\ 0 & \text{if} \quad r > 1, \end{cases}$$

so $\alpha = \binom{5}{0}\left(\frac{1}{2}\right)^5 + \frac{1}{5}\binom{5}{1}\left(\frac{1}{2}\right)^5 = \frac{1}{16}$.

Now suppose the trial was conducted until we got r heads which took n trials (random). Then

$$P(n|r) = \binom{n-1}{r-1} p^r (1-p)^{n-r} \quad n = r, r+1, \dots \ .$$

Note $L_B(p|r) = L_{NB}(p|n) = p^r(1-p)^{n-r}$, that is, the binomial and negative binomial likelihoods are the same.

Now suppose for $\alpha = 1/16$ and $r = 1$, so

$$P\left(N = n|r = 1, p = \frac{1}{2}\right) = \left(\frac{1}{2}\right)^n,$$

such that

$$P(N \leq 4) = \frac{1}{2} + \frac{1}{4} + \frac{1}{8} + \frac{1}{10} = \frac{15}{16}$$

or $P(N \geq 5) = 1/16$, hence

$$T(n) = \begin{cases} 1 & \text{if} \quad n \geq 5 \\ 0 & \text{if} \quad n \leq 4 \end{cases}.$$

Now if the data in both experiments were $r = 1$, $n = 5$ then for the Binomial trials we would reject with probability $\frac{1}{5}$ and with the negative binomial trials we would reject with probability 1. So as previously noted the likelihood principle and likelihood tests are contradicted although in both cases

$$L = p(1-p)^4.$$

Clearly the difference is due to differing rejection regions and that has to do with fixing the same α in both experiments and maximizing $1 - \beta$ or minimizing β.

Suppose instead we sought in each experiment to minimize $\alpha + k\beta$ where k is the relative importance of β with respect to α recalling $\alpha = P(\text{reject } H_0|H_0)$, $\beta = P(\text{reject } H_1|H_1)$. For the simple dichotomy where $f_0(D)/f_1(D)$ represents the ratio of likelihoods,

$$\alpha + k\beta = \alpha + k[1 - (1 - \beta)]$$

$$= \int_s T(D)f_0(D)d\mu + k - k\int_s T(D)f_1(D)d\mu$$

$$= k + \int T(D)[f_0 - kf_1]d\mu.$$

Minimization of the above occurs whenever

$$f_0(D) - kf_1(D) > 0, \quad T(D) = 0$$
$$f_0(D) - kf_1(D) < 0, \quad T(D) = 1$$

and $T(D)$ is arbitrary when $f_0(D) = kf_1(D)$. Then the test $T(D)$ that minimizes $\alpha + k\beta$ rejects H_0 when

$$\frac{f_0(D)}{f_1(D)} < k$$

and is a likelihood test. Note it doesn't fix α in advance and for each n, $\alpha + k\beta$ will be minimized and a function of n.

Now in the case discussed the Likelihood test was the same for r and n whether it was L_B or L_{NB}. In the particular case $H_0 : p = \frac{1}{2}, H_1 : p = p_1 < \frac{1}{2}$ we reject if

$$\frac{(1/2)^r(1/2)^{n-r}}{p_1^r(1 - p_1)^{n-r}} < k.$$

Solving for r yields rejection if

$$r < \frac{\log k}{\log\left(\dfrac{1 - p_1}{p_1}\right)} + \frac{n\log\left(1 - \dfrac{p_1}{1/2}\right)}{\log\left(\dfrac{1 - p_1}{1/2} \times \dfrac{1/2}{p_1}\right)} = \frac{\log k}{\log\left(\dfrac{1 - p_1}{p_1}\right)} + na_1(p_1).$$

Now for $p_1 < \frac{1}{2}$

$$\frac{1 - p_1}{1/2} < \frac{1/2}{p_1} \quad \text{since} \quad (1 - p_1)p_1 < \frac{1}{4}$$

then $a_1(p_1) < \frac{1}{2}$ so that

$$r < \frac{\log k}{\log\left(\dfrac{1 - p_1}{p_1}\right)} + a_1(p_1)n \quad \text{is the likelihood test.}$$

But the rejection region for the LRT depends on fixing α for each n and calculating

$$r < \frac{\log k_\alpha(n)}{\log\left(\dfrac{1-p_1}{p_1}\right)} + a_1(p_1)n,$$

such that

$$P\left[r < \frac{\log k_\alpha(n)}{\log\left(\dfrac{1-p_1}{p_1}\right)} + a_1(p_1)n\,\middle|\,H_0\right] = \alpha.$$

Suppose for large samples we use the normal approximation to the binomial then

$$P\left[\frac{r - np}{\sqrt{npq}} < -z_\alpha\right] = \alpha \quad \text{where} \quad \int_{-\infty}^{-z_\alpha} \frac{1}{\sqrt{2\pi}} e^{-z^2/2}\,dz = \alpha.$$

The value of $\log k_\alpha(n)$ under $H_0 : p = \frac{1}{2}$ can be obtained. Since

$$P\left(\frac{r - \dfrac{n}{2}}{\frac{1}{2}\sqrt{n}} < \frac{\log k_\alpha(n)}{\frac{1}{2}\sqrt{n}\log\left(\dfrac{1-p_1}{p_1}\right)} - \frac{(\frac{1}{2} - a_1(p_1))n}{\frac{1}{2}\sqrt{n}}\right) = \alpha,$$

then

$$-z_\alpha \doteq \frac{\log k_\alpha(n)}{\frac{1}{2}\sqrt{n}\log\left(\dfrac{1-p_1}{p_1}\right)} - \frac{(\frac{1}{2} - a_1(p_1))n}{\frac{1}{2}\sqrt{n}}$$

and

$$-\frac{1}{2}\sqrt{n}\, z_\alpha + \left(\frac{1}{2} - a_1(p_1)\right)n \doteq \log k_\alpha(n)\,\Big/\,\log\left(\frac{1-p_1}{p_1}\right).$$

Hence

$$P\left[r < \frac{\log k_\alpha(n)}{\log\dfrac{1-p_1}{p_1}} + \frac{n\log\left(\dfrac{1-p_1}{1/2}\right)}{\log\left(\dfrac{1-p_1}{p_1}\right)}\right] \doteq P\left[r < -\frac{1}{2}\sqrt{n}\, z_\alpha + \frac{n}{2}\right] = \alpha.$$

What happens if one continues sampling? Is one certain to find that for some n

$$r < \frac{n}{2} - \frac{1}{2}z_\alpha\sqrt{n} \quad \text{when } H_0 \text{ is true}$$

that is, be sure to reject H_0 when H_0 is true? We can answer this by appealing to the Law of the Iterated Logarithm.

5.9   LAW OF THE ITERATED LOGARITHM (LIL)

Let $X_1, X_2, \ldots$ be i.i.d. random variables with $E(X_i) = \mu$ $Var(X_i) = \sigma^2$ and $E|X|^{2+\delta} < \infty$ for some $\delta > 0$. Then with probability 1 the inequality

$$\sum_{i=1}^{n} X_i < n\mu - (n\sigma^2 \lambda \log \log n)^{\frac{1}{2}} \quad \text{or} \quad \frac{\Sigma X_i - n\mu}{\sigma \sqrt{n}} < -(\lambda \log \log n)^{\frac{1}{2}}$$

is satisfied for infinitely many n if $\lambda < 2$ but for only finitely many if $\lambda > 2$. Similarly

$$\Sigma X_i > n\mu + (n\sigma^2 \lambda \log \log n)^{\frac{1}{2}} \quad \text{or} \quad \frac{\Sigma X_i - n\mu}{\sigma \sqrt{n}} > (\lambda \log \log n)^{\frac{1}{2}}$$

is satisfied for infinitely many n if $\lambda < 2$, but only for finitely many n if $\lambda > 2$. Further (almost surely)

$$\limsup_{n \to \infty} \frac{(\Sigma X_i - n\mu)/\sigma\sqrt{n}}{(2 \log \log n)^{\frac{1}{2}}} = 1.$$

We apply the LIL to n i.i.d. binary variables where $P(X_i = 0) = P(X_i = 1) = \frac{1}{2}$ where $E(X_i) = \frac{1}{2}$, $\operatorname{var}(X_i) = \frac{1}{4}$. Then for $\lambda = 1$, consider the event

$$r < \frac{n}{2} - \left(\frac{n}{4} \log \log n\right)^{\frac{1}{2}}.$$

Sooner or later if we continue sampling then for sufficiently large n

$$(\log \log n)^{\frac{1}{2}} > z_\alpha$$

since z_α is a constant. Then

$$-\left(\frac{n}{4} \log \log n\right)^{\frac{1}{2}} < -\frac{\sqrt{n}}{2} z_\alpha$$

and

$$\frac{n}{2} - \left(\frac{n}{4} \log \log n\right)^{\frac{1}{2}} < \frac{n}{2} - \frac{\sqrt{n}}{2} z_\alpha.$$

Therefore, with probability 1 the inequality

$$r < \frac{n}{2} - \frac{\sqrt{n}}{2} z_\alpha$$

is satisfied for infinitely many n and one is therefore certain to reject H_0 when H_0 is true if one sets out to do this by continually sampling until one finds

$$r < \frac{n}{2} - \frac{\sqrt{n}}{2} z_\alpha,$$

which is certain to happen sooner or later. In other words, we have sampled to a foregone conclusion.

Now consider the likelihood test which is reject H_0 if for *fixed k*

$$r < \frac{\log k}{\log\left(\dfrac{1-p_1}{p_1}\right)} + na_1(p_1) = \frac{n}{2} + \frac{\log k}{\log\left(\dfrac{1-p_1}{p_1}\right)} - (1/2 - a_1(p_1))n$$

$$0 < a_1(p_1) < \frac{1}{2}.$$

Now for sufficiently large n and $\lambda = 4$,

$$-\frac{\log k}{\log\left(\dfrac{1-p_1}{p_1}\right)} + (1/2 - a_1(p_1))n < (n \log \log n)^{\frac{1}{2}}$$

or

$$\frac{n}{2} + \frac{\log k}{\log\left(\dfrac{1-p_1}{p_1}\right)} - (1/2 - a_1(p_1))n > \frac{n}{2} - (n \log \log n)^{\frac{1}{2}}$$

is satisfied for only finitely many n. Moreover, $r < n/2 - (n \log \log n)^{\frac{1}{2}}$ for only finitely many n with probability one. Therefore it is not certain that using the likelihood test by one who unscrupulously sets out to reject the null hypothesis will do so. One could very well sample forever and still not reject. But using the N-P approach for each n will make rejection certain.

A bound on $P(\text{reject } H_0|H_0)$ for all n can be obtained for the likelihood test

$$\frac{L(\theta_0)}{L(\theta_1)} < k < 1.$$

Let $f_{0m} = f(x_1, \ldots, x_m|\theta_0)$ and $f_{1m} = f_{1m}(x_1, \ldots, x_m|\theta_1)$ and

$$Q_m = \frac{f_{0m}}{f_{1m}} < k \text{ reject } H_0 : \theta = \theta_0.$$

On the nth trial the probability of rejecting H_0 when H_0 is true is

$$P_{0n} = P[Q_m \geq k \text{ for } m = 1, \ldots, n-1 \text{ and } Q_n < k],$$

such that

$$P(\text{reject } H_0|H_0) = \sum_{n=1}^{\infty} P_{0n} = \sum_{n=1}^{\infty} \int_{W_{1n}} f_{0n} dx^{(n)},$$

where $x^{(n)} = (x_1, \ldots, x_n)$ and

$$W_{1n} = [(x_1, \ldots, x_m); \quad Q_m \geq k \text{ for } m = 1, \ldots, n-1, \ Q_n < k].$$

For the test to terminate on the nth observation we need $f_{0n} < kf_{1n}$ or

$$\int_{W_{1n}} f_{0n} dx^{(n)} \leq k \int_{W_{1n}} f_{1n} dx^{(n)}.$$

Now

$$P(\text{reject } H_0|H_0) = \sum_{n=1}^{\infty} \int_{W_{1n}} f_{0n} dx^{(n)} \leq k \sum_{n=1}^{\infty} \int_{W_{1n}} f_{1n} dx^{(n)}.$$

Also $\int_{W_{1n}} f_{1n} dx^{(n)} = P(\text{reject } H_0 \text{ on } n\text{th observation } |H_1)$ such that

$$\sum_{n=1}^{\infty} \int_{W_{1n}} f_{1n} dx = P[\text{reject } H_0|H_1].$$

Hence $P(\text{rejecting } H_0|H_0) \leq kP(\text{reject } H_0|H_1] \leq k$ such that probability that the test never terminates is greater than $1 - k$ which can be crude in many cases.

An application of LIL was given by Barnard (1969).

Example 5.13

In British Standards 3704, 1964 on Condoms the sampling clause reads as follows:

3.a. Sampling. "Specimens constituting the test sample shall be taken at random from each quantum of production. The number of these specimens shall not be less than 1% of the number of articles in each quantum. ...."

The number of test specimens n and the number of rejected specimens r from a sequence of production quanta shall be recorded. The cumulative total of test specimens N and the cumulative total of rejects R shall be recorded and the products shall be deemed to comply with the requirements of this British Standard or acceptance of $p < 0.01$ if

$$R \leq 0.01N + 3\sqrt{0.01N}$$

or introducing the standardized variable

$$Z_N = \frac{R - Np}{\sqrt{Np(1-p)}} \le \frac{(0.01 - p)\sqrt{N}}{\sqrt{p(1-p)}} + \frac{0.3}{\sqrt{p(1-p)}}.$$

Consider $H_0 : p < 0.01$ vs. $H_1 : p \ge 0.01$.
   Now for $p \ge 0.01$ and for $a > 0$

$$Z_N \le \frac{-a\sqrt{N}}{\sqrt{p(1-p)}} + \frac{0.3}{\sqrt{p(1-p)}}.$$

Since this right-hand side declines faster than $-\sqrt{2\log\log N}$ then Z_N, sooner or later, will be greater than the right-hand side with probability 1. It is also true if $p = 0.01$ where the right-hand side is constant. So if H_0 is false it will be rejected sooner or later.
   Now if $p < 0.01$

$$Z_N < \frac{b\sqrt{N}}{\sqrt{p(1-p)}} + \frac{0.3}{\sqrt{p(1-p)}} \quad b > 0.$$

Since the right-hand side grows faster than $(2\log\log N)^{\frac{1}{2}}$ this is satisfied for all but finitely many N so there is a non-zero probability that this will be satisfied for all N.
   The procedure then is to continue sampling and accepting until $R > 0.01N + 3\sqrt{0.01N}$. Then reject and reset the manufacturing device. At any rate this is a test whose power is virtually 1.

5.10   SEQUENTIAL ANALYSIS

Up to now we have studied cases where the sample size N was a fixed value determined prior to taking observations except in negative binomial and multinomial sampling where the sample size N was a random variable. We now study the situation where we sample sequentially to test a null hypothesis versus a rival hypothesis. Here the test may terminate at some value of $N = n$ as opposed to waiting until a fixed number of observations are in hand.

5.11   SEQUENTIAL PROBABILITY RATIO TEST (SPRT)

For testing a simple H_0 versus a simple alternative H_1, where

$$H_0 : f = f_{\theta_0} \quad \text{versus} \quad H_1 : f = f_{\theta_1},$$

we define the ratio

$$\frac{\prod_{i=1}^{n} f(x_i|\theta_0)}{\prod_{i=1}^{n} f(x_i|\theta_1)} = \frac{f_{0n}}{f_{1n}} = Q_n, \quad n = 1, 2, \ldots$$

and two positive constants A and B.

The test as defined by Wald (1947) at each n accepts H_0 if $Q_n \geq B$, accepts H_1 if $Q_n \leq A$ and requires sampling another observation if

$$A < Q_n < B.$$

A and B are to be determined such that the test is of size α and power $1 - \beta$. Set

$$P_{1N} = P[A < Q_n < B, \quad n = 1, \ldots, N - 1, \quad \text{and} \quad Q_N \geq B | H_1],$$
$$W_{0N} = [(x_1, \ldots, x_N) : A < Q_n < B, \quad n = 1, \ldots, N - 1, \quad Q_N \geq B],$$

then

$$P_{1N} = \int_{W_{0N}} f_{1N} d\mu.$$

Now the probability of accepting H_0 when H_1 is true is

$$\beta = \sum_{N=1}^{\infty} P_{1N}.$$

Similarly,

$$P_{0N} = P[A < Q_n < B : n = 1, \ldots, N - 1 \quad \text{and} \quad Q_N \leq A | H_0],$$
$$W_{1N} = [(x_1, \ldots, x_N) : A < Q_n < B, \quad = 1, \ldots, N - 1, \quad Q_N \leq A],$$

$$P_{0N} = \int_{W_{1N}} f_{0N} d\mu,$$

and the probability of accepting H_1 where H_0 is true is

$$\alpha = \sum_{N=1}^{\infty} P_{0N}.$$

Now Wald showed that the test terminates with probability one that is, sooner or later the boundary points will be hit. Thus for any given sample size N that terminates with acceptance of H_0 we have

$$f_{0N} \geq B f_{1N}.$$

Integrating both sides over W_{0N} and then summing with respect to N yields

$$1 - \alpha \geq B\beta,$$

or

$$\frac{1}{B} \geq \frac{\beta}{1 - \alpha} \geq \beta.$$

Similarly, for given sample size N that terminates with acceptance of H_1,

$$\alpha \leq A(1 - \beta) \leq A,$$

resulting in reasonable upper bounds on α and β when

$$A < 1 < B.$$

Now suppose that the true errors incurred, by using

$$\frac{\alpha}{1 - \beta} = A \quad \text{and} \quad \frac{\beta}{1 - \alpha} = \frac{1}{B},$$

are α' and β'. By our previous work the true α' and β' are such that

$$\frac{\alpha'}{1 - \beta'} \leq \frac{\alpha}{1 - \beta} = A; \quad \frac{\beta'}{1 - \alpha'} \leq \frac{\beta}{1 - \alpha} = \frac{1}{B}$$

or

$$\alpha'(1 - \beta) \leq \alpha(1 - \beta'); \quad \beta'(1 - \alpha) \leq \beta(1 - \alpha').$$

This results in

$$\alpha' + \beta' \leq \alpha + \beta,$$

implying that either $\alpha' \leq \alpha$ or $\beta' \leq \beta$ or both. Hence the overall protection $\alpha' + \beta'$ is at least as good as the presumed protection $\alpha + \beta$ derived from

$$\frac{\alpha}{1 - \beta} = A \quad \text{and} \quad \frac{\beta}{1 - \alpha} = \frac{1}{B}.$$

One of the features of this SPRT is that for fixed α and β the SPRT minimizes the expected sample sizes so on average the SPRT will come to a conclusion more quickly than the fixed sample size test.

Notice also that this is basically a likelihood test that conforms to Neyman-Pearson size and power modifications.

REFERENCES

Barnard, G. A. (1969). Practical application of tests of power one. *Bulletin of the International Statistical Institute*, **XLIII**, **1**, 389–393.

Hacking, I. (1965). *Logic of Statistical Inference.* Cambridge: Cambridge University Press.

Lehmann, E. L. (1950). Some principles of the theory of testing hypothesis. *Annals of Mathematical Statistics*, **21**, 1–26.

Lehmann, E.L. (1959). *Testing Statistical Hypothesis.* New York: Wiley.

Neyman, J. and Pearson, E. S. (1936–1938). Contributions to the theory of testing statistical hypotheses, *Statistical Research Memoirs*, **I**, 1–37; **II**, 25–57.

Wald, A. (1947). *Sequential Analysis.* New York: Wiley.

CHAPTER SIX

Elements of Bayesianism

Bayesian inference involves placing a probability distribution on all unknown quantities in a statistical problem. So in addition to a probability model for the data, probability is also specified for any unknown parameters associated with it. If future observables are to be predicted, probability is posed for these as well. The Bayesian inference is thus the conditional distribution of unknown parameters (and/or future observables) given the data. This can be quite simple if everything being modeled is discrete, or quite complex if the class of models considered for the data is broad.

This chapter presents the fundamental elements of Bayesian testing for simple versus simple, composite versus composite and for point null versus composite alternative, hypotheses. Several applications are given. The use of Jeffreys "noninformative" priors for making general Bayesian inferences in binomial and negative binomial sampling, and methods for hypergeometric and negative hypergeometric sampling, are also discussed. This discussion leads to a presentation and proof of de Finetti's theorem for Bernoulli trials. The chapter then gives a presentation of another de Finetti result that Bayesian assignment of probabilities is both necessary and sufficient for "coherence," in a particular setting. The chapter concludes with a discussion and illustration of model selection.

6.1 BAYESIAN TESTING

Recall

$$P(A|B)P(B) = P(A, \ B) = P(B|A)P(A),$$

where A and B are events in S. If X, Y are jointly discrete

$$f(x, \ y) = P(X = x, \ Y = y) = P(X = x|Y = y)P(Y = y) = P(Y = y|X = x)P(X = x).$$

Modes of Parametric Statistical Inference, by Seymour Geisser
Copyright © 2006 John Wiley & Sons, Inc.

If X and Y are jointly absolutely continuous so that

$$F_{X,Y}(x, \ y) = \int_{-\infty}^{y} \int_{-\infty}^{x} f_{X,Y}(u, \ v) du \ dv$$

then

$$f(x,y) = f(x|y) \ f(y) = f(\ y|x) f(x).$$

In general, whether X, Y are discrete, continuous or one is discrete and the other continuous

$$f(y|x) = \frac{f(x|y) \ f(y)}{f(x)}$$

$$f(x) = \begin{cases} \sum_{y} f(x|y) \ f(y) & \text{if } Y \text{ is discrete} \\ \int f(x|y) \ f(y) dy & \text{if } Y \text{ is continuous} \end{cases}$$

Note also that

$$f(y|x) \propto f(x|y) f(y).$$

In a situation where we have $D \sim f(D|\theta)$ and an assumed prior probability function for $\theta \in \Theta$ namely $g(\theta)$ then

$$f(\theta|D) \propto L(\theta|D)g(\theta).$$

Now suppose that $H_0 : \theta = \theta_0$ vs. $H_1 : \theta = \theta_1$. If we assume that $P(\theta = \theta_0) = p$ and $P(\theta = \theta_1) = 1 - p$ then

$$P(\theta = \theta_0|D) \propto L(\theta_0|D)p,$$
$$P(\theta = \theta_1|D) \propto L(\theta_1|D)(1 - p),$$

and

$$\frac{P(\theta = \theta_0|D)}{P(\theta = \theta_1|D)} = \frac{p}{1 - p} \ \frac{L(\theta_0|D)}{L(\theta_1|D)}.$$

So we have that the Posterior Odds = Prior odds $\times$ Likelihood Ratio or

Log Posterior odds $-$ log Prior odds = log Likelihood Ratio, where the right-hand side can be considered as the information in the experiment relative to the two hypotheses or how you modify the prior odds by data D. If $(L(\theta_0|D))/(L(\theta_1|D)) = 1$ the data are uniformative as to discriminating between hypotheses.

Example 6.1

Suppose $L(\theta|r) = \theta^r(1 - \theta)^{n-r}$ then

$$\frac{P(\theta = \theta_0|r)}{P(\theta = \theta_1|r)} = \frac{\theta_0^r(1 - \theta_0)^{n-r}}{\theta_1^r(1 - \theta_1)^{n-r}} \times \frac{p}{1 - p}.$$

If $X_1, \ldots, X_n$ are i.i.d. then

$$L(\theta|D) = \prod_{i=1}^{n} f(x_i|\theta),$$

and thus

$$P(\theta = \theta_0|x_1, \ldots, x_n) = \frac{p \prod_{i=1}^{n} f(x_i|\theta_0)}{p \prod_{i=1}^{n} f(x_i|\theta_0) + (1 - p) \prod_{i=1}^{n} f(x_i|\theta_1)}$$

$$= \frac{1}{1 + \frac{(1 - p)}{p} \prod_{i=1}^{n} \frac{f(x_i|\theta_1)}{f(x_i|\theta_0)}},$$

$$P(\theta = \theta_1|x_1, \ldots, x_n) = \frac{1}{1 + \frac{p}{1 - p} \prod_{i=1}^{n} \frac{f(x_i|\theta_0)}{f(x_i|\theta_1)}}.$$

Now suppose θ_1 is actually the true value. We now will show that the limiting posterior probability, $\lim_{n \to \infty} P(\theta = \theta_1|x_1, \ldots, x_n) = 1$, if θ_1 is the true value.
    Consider

$$E_{\theta_1}\left(\log \frac{f(x_i|\theta_0)}{f(x_i|\theta_1)}\right) = E_{\theta_1}\left(\log \frac{f_0}{f_1}\right) = \mu, \quad \text{say.}$$

Jensen's inequality states that $E[g(Y)] \geq g(E(Y))$ if $g(Y)$ is a convex function with equality only if Y is degenerate. Now $-\log Y$ is convex since $-\log Y$ is twice differentiable and $g''(y) \geq 0$ implies convexity, that is, $g(\alpha z + \beta y) \leq \alpha g(z) + \beta g(y)$ for all z and y and $\alpha + \beta = 1$ for $\alpha > 0$ and $\beta > 0$.
    Hence

$$-\mu = E_{\theta_1}\left(-\log \frac{f_0}{f_1}\right) > -\log E_{\theta_1}\left(\frac{f_0}{f_1}\right),$$

or

$$\mu < \log E_{\theta_1}\left(\frac{f_0}{f_1}\right) = \log \int \frac{f_0}{f_1} f_1 dx_i = \log 1 = 0.$$

Since $Y_i = \log\frac{f(x_i|\theta_0)}{f(x_i|\theta_1)}$ are i.i.d. and $E_{\theta_1}(Y_i) = \mu$,

$$\lim_{n\to\infty} \frac{1}{n}\sum_{i=1}^{n} \log\frac{f(x_i|\theta_0)}{f(x_i|\theta_1)} = \mu$$

by the Strong Law of Large Numbers (SLLN).

SLLN: Let $Y_1, Y_2, \ldots$ be a sequence of random variables i.i.d. with mean μ and for every pair $\epsilon > 0$ and $\delta > 0$ there is an N such that

$$P[\sup_{n\geq N} |\bar{Y}_n - \mu| < \epsilon] \geq 1 - \delta \quad \text{i.e. with probability 1.}$$

Now

$$\frac{1}{n}\log \prod_{i=1}^{n} \frac{f(x_i|\theta_0)}{f(x_i|\theta_1)} = \frac{1}{n}\sum_{i=1}^{n} \log\frac{f(x_i|\theta_0)}{f(x_i|\theta_1)} \longrightarrow \mu < 0,$$

so that

$$\lim_{n\to\infty} \prod_{1}^{n} \frac{f(x_i|\theta_0)}{f(x_i|\theta_1)} = \lim_{n\to\infty} e^{n\mu} \longrightarrow 0 \quad \text{since } \mu < 0.$$

Hence with probability 1

$$\lim_{n\to\infty} P(\theta = \theta_1|x_1, \ldots, x_n) = 1 \text{ if } \theta_1 \text{ is true.}$$

6.2   TESTING A COMPOSITE VS. A COMPOSITE

Suppose we can assume that the parameter or set of parameters $\theta \in \Theta$ is assigned a prior distribution (subjective or objective) that is, we may be sampling θ from a hypothetical population specifying a probability function $g(\theta)$, or our beliefs about θ can be summarized by a $g(\theta)$ or we may assume a $g(\theta)$ that purports to reflect our prior ignorance. We are interested in deciding whether

$$H_0 : \theta \in \Theta_0 \quad \text{or} \quad H_1 : \theta \in \Theta_1 \quad \text{for} \quad \Theta_0 \cap \Theta_1 = \emptyset.$$

Now the posterior probability function is

$$p(\theta|D) \propto L(\theta|D)g(\theta) \quad \text{or} \quad p(\theta|D) = \frac{L(\theta|D)g(\theta)}{\int_{\Theta} L(\theta|D)dG(\theta)},$$

using the Lebesgue-Stieltjes integral representation in the denominator above. Usually $\Theta_0 \cup \Theta_1 = \Theta$ but this is not necessary. We calculate

$$P(\theta \in \Theta_0|D) = \int_{\Theta_0} dP(\theta|D)$$

$$P(\theta \in \Theta_1|D) = \int_{\Theta_1} dP(\theta|D)$$

and calculate the posterior odds

$$\frac{P(\theta \in \Theta_0|D)}{P(\theta \in \Theta_1|D)}$$

and if this is greater than some predetermined value k choose H_0, and if less choose H_1, and if equal to k be indifferent.

Example 6.2

Suppose

$$L(\theta|D) = \theta^r(1 - \theta)^{n-r}, \quad 0 < \theta < 1,$$

and we assume $g(\theta) = 1, 0 < \theta < 1$ and

$$H_0 : \theta < \frac{1}{2}, \qquad H_1 : \theta \geq \frac{1}{2}.$$

Then

$$p(\theta|r) \propto \theta^r(1 - \theta)^{n-r}$$

and

$$Q = \frac{P(\theta < \frac{1}{2}|r)}{P(\theta \geq \frac{1}{2}|r)} = \frac{\int_0^{\frac{1}{2}} \theta^r(1 - \theta)^{n-r}d\theta}{\int_{\frac{1}{2}}^1 \theta^r(1 - \theta)^{n-r}d\theta}.$$

We show that

$$P(\theta < 1/2|r) = P\left(X < \frac{n - r + 1}{r + 1}\right),$$

where $X \sim F(2r+2, \ 2n-2r+2)$, an F distribution with the arguments as the degrees of freedom, so that

$$Q = P\left(X < \frac{n-r+1}{r+1}\right) \Big/ P\left(X \geq \frac{n-r+1}{r+1}\right).$$

Let $\gamma = \dfrac{\theta}{1-\theta}$, $d\gamma = \dfrac{1}{(1-\theta)^2}$, $\theta = \dfrac{\gamma}{1+\gamma}$ and $1-\theta = \dfrac{1}{1+\gamma}$, so that

$$p(\gamma|r) \propto \frac{\gamma^r}{(1+\gamma)^r} \times \frac{1}{(1+\gamma)^{n-r}} \times \frac{1}{(1+\gamma)^2} = \frac{\gamma^r}{(1+\gamma)^{n+2}}.$$

Then it is clear that

$$X = \frac{(n-r+1)}{(r+1)} \frac{\theta}{1-\theta} \sim F(2r+2, \ 2n-2r+2).$$

Hence

$$P(\theta < 1/2) = P(2\theta < 1) = P(\theta < 1-\theta) = P\left(\frac{\theta}{1-\theta} < 1\right)$$

$$= P\left(\frac{n-r+1}{r+1} \cdot \frac{\theta}{1-\theta} < \frac{n-r+1}{r+1}\right) = P\left(X < \frac{n-r+1}{r+1}\right)$$

and

$$Q = \frac{P\left(X < \dfrac{n-r+1}{r+1}\right)}{1 - P\left(X < \dfrac{n-r+1}{r+1}\right)}.$$

To test,

$$H_0 \ : \ \theta = \theta_0 \quad \text{vs.} \quad H_1 \ : \ \theta \neq \theta_0,$$

the previous setup for a composite will not do. Also picking a null $H_0 : \theta = \theta_0$ would imply that this may be an important value and that $\theta = \theta_0$ is quite likely.

Let $P(\theta = \theta_0) = p$ and assume that all other values are subject to a prior density $g(\theta) \ \theta \in \Theta$ except $\theta = \theta_0$. Observe that the prior CDF satisfies

$$G(\theta) = \begin{cases} \int_{-\infty}^{\theta} dG(\theta') & \theta < \theta_0 \\ \int_{-\infty}^{\theta_0-} dG(\theta') + p & \theta = \theta_0 \\ \int_{-\infty}^{\theta_0-} dG(\theta') + p + \int_{\theta_0}^{\theta} dG(\theta') & \theta > \theta_0. \end{cases}$$

Then

$$P(\theta = \theta_0|D) = pf(D|\theta_0) \Big/ \left[\int_{-\infty}^{\theta_0^-} f(D|\theta)dG + f(D|\theta_0)p + \int_{\theta_0}^{\infty} f(D|\theta_0)dG \right].$$

Suppose $G(\theta)$ has a continuous derivative $(dG)/(d\theta) = g(\theta)$ everywhere except at $\theta = \theta_0$. Then $P(\theta = \theta_0|D) \propto pf(D|\theta_0)$ and

$$P(\theta \ne \theta_0|D) \propto \int_\theta f(D|\theta)g(\theta)d\theta, \quad \text{where} \quad \int_\Theta g(\theta)d\theta = 1 - p \quad \text{for} \quad \theta \ne \theta_0.$$

Example 6.3

Let $X_1, \ldots, X_n$ be i.i.d. Bernoulli variates with $E(X_i) = \theta$. Consider testing

$$H_0 \;:\; \theta = \theta_0.$$

Let

$$G(\theta) = \begin{cases} \theta(1 - p) & 0 \le \theta < \theta_0 \\ \theta_0(1 - p) + p & \theta = \theta_0 \\ p + \theta(1 - p) & \theta_0 < \theta \le 1 \end{cases}$$

Note

$$\frac{dG(\theta)}{d\theta} = (1 - p) \quad \forall \quad \theta \ne \theta_0.$$

Then $g(\theta)$ is uniform over all other values and

$$P(\theta = \theta_0|D) \propto p\theta_0^r(1 - \theta_0)^{n-r}$$

$$P(\theta \ne \theta_0|D) \propto \int_0^1 \theta^r(1 - \theta)^{n-r}(1 - p)d\theta \quad \text{for} \quad \theta \ne \theta_0$$

and

$$Q = \frac{P(\theta = \theta_0|D)}{P(\theta \ne \theta_0|D)} = \frac{p\theta_0^r(1 - \theta_0)^{n-r}}{(1 - p)\int_0^1 \theta^r(1 - \theta)^{n-r}d\theta} = \frac{p\theta_0^r(1 - \theta_0)^{n-r}}{(1 - p)\dfrac{\Gamma(r + 1)\Gamma(n - r + 1)}{\Gamma(n + 2)}}$$

$$= \frac{p}{(1 - p)}(n + 1)\binom{n}{r}\theta_0^r(1 - \theta_0)^{n-r} = \frac{p}{(1 - p)}(n + 1)P[r|n, \theta_0].$$

For large n using the normal approximation to the binomial

$$Q \approx \frac{p}{1-p}(n+1) \cdot \frac{1}{\sqrt{2\pi n \theta_0 (1-\theta_0)}} e^{-\frac{1}{2}\frac{(r-n\theta_0)^2}{n\theta_0(1-\theta_0)}}$$

$$\approx \frac{p}{1-p}\sqrt{\frac{n}{2\pi\theta_0(1-\theta_0)}}e^{-\frac{1}{2}\frac{(r-n\theta_0)^2}{n\theta_0(1-\theta_0)}}.$$

Suppose $\theta_0 = \frac{1}{2}$ and $p = \frac{1}{2}$ and we are testing for bias in a coin. Then let

$$Q_{ht} = (n+1)\binom{n}{r}\left(\frac{1}{2}\right)^n,$$

where $h = \#$ heads and $t = \#$ tails. When $n = 1$, $r = 0$ or 1 and

$$Q_{10} = Q_{01} = 1.$$

Is this sensible? Since it is no different than the prior odds $\frac{1}{2} \div \frac{1}{2} = 1$, there is no information in the first observation (it has to be heads or tails). Further suppose n is even and $n = r = n - r = t$ that is, equal number of successes and failures so $n = 2r$. Then

$$Q_{r,r} = \frac{(n+1)!}{2^n r!(n-r)!} = \frac{(2r+1)!}{2^{2r} r! r!}.$$

Suppose another observation is taken and is heads, then

$$Q_{r+1,\ r} = \frac{(2r+2)!}{2^{2r+1}(r+1)! r!}$$

or if it is tails

$$Q_{r,r+1} = \frac{(2r+2)!}{2^{2r+1} r!(r+1)!}$$

so $Q_{r+1,r} = Q_{r,r+1}$. Further,

$$Q_{r,r} = \frac{(2r+2)}{(2r+2)} \frac{(2r+1)!}{2^{2r} r! r!} = \frac{(2r+2)!}{2^{2r+1}(r+1)! r!} = Q_{r+1,\ r} = Q_{r,\ r+1}.$$

Therefore the odds don't change after an observation is made when previously we had an equal number of heads and tails. Again the interpretation is that the next toss has to be a head or a tail and this does not provide any new information. This is eminently reasonable according to Jeffreys (1961), who proposed this test.

Example 6.4

Test of equality of two binomial parameters.
    Consider the sampling situation

$$R_1 \sim Bin(\theta_1,\, n_1) \quad R_2 \sim Bin(\theta_2,\, n_2)$$

$$p(r_i|\theta_i) = \binom{n_i}{r_i} \theta_i^{r_i}(1-\theta_i)^{n_i-r_i}, \quad i = 1,\, 2$$

$$H_0 : \theta_1 = \theta_2 \quad \text{vs} \quad H_1 : \theta_1 \neq \theta_2$$

Assumptions *a priori* are

$$g(\theta_1,\, \theta_2) = (1-p) \quad \text{for} \quad \theta_1 \neq \theta_2,$$
$$g(\theta) = p \quad \text{for} \quad \theta_1 = \theta_2 = \theta.$$

Therefore,

$$P(\theta_1 = \theta_2 | r_1,\, r_2) \propto p \int_0^1 \theta^{r_1+r_2}(1-\theta)^{n_1+n_2-r_1-r_2} d\theta$$

$$= p \frac{(r_1+r_2)!(n_1+n_2-r_1-r_2)!}{(n_1+n_2+1)!},$$

$$P(\theta_1 \neq \theta_2 | r_1,\, r_2) \propto (1-p) \int_0^1 \int_0^1 \prod_{i=1}^2 \theta_i^{r_i}(1-\theta_i)^{n_i-r_i} d\theta_1 d\theta_2$$

$$= \frac{(1-p)r_1!(n_1-r_1)!r_2!(n_2-r_2)!}{(n_1+1)!(n_2+1)!}.$$

Then

$$Q = \frac{P(\theta_1 = \theta_2 | r_1,\, r_2)}{P(\theta_1 \neq \theta_2 | r_1,\, r_2)} = \frac{p}{1-p} \times \frac{r!(n-r)!(n_1+1)!(n_2+1)!}{(n+1)!r_1!(n_1-r_1)!r_2!(n_2-r_2)!}.$$

For $r_1 + r_2 = r$, $n_1 + n_2 = n$

$$Q = \frac{p}{1-p} \times \frac{(n_1+1)(n_2+1)}{(n+1)} \frac{\binom{n_1}{r_1}\binom{n_2}{r_2}}{\binom{n}{r}}.$$

If $p = \frac{1}{2}$, using the binomial approximation to the hypergeometric

$$Q \approx \frac{(n_1 + 1)(n_2 + 1)}{(n + 1)} \binom{r}{r_1} \left(\frac{n_1}{n}\right)^{r_1} \left(\frac{n_2}{n}\right)^{r_2}.$$

6.3 SOME REMARKS ON PRIORS FOR THE BINOMIAL

1. *Personal Priors.* If one can elicit a personal subjective prior for θ then his posterior for θ is personal as well and depending on the reasoning that went into it may or may not convince anyone else about the posterior on θ. A convenient prior that is often used when subjective opinion can be molded into this prior is the beta prior

$$g(\theta|a, b) \propto \theta^{a-1}(1 - \theta)^{b-1},$$

when this is combined with the likelihood to yield

$$g(\theta|a, b, r) \propto \theta^{a+r-1}(1 - \theta)^{b+n-r-1}.$$

2. *So-Called Ignorance or Informationless or Reference Priors.* It appears that in absence of information regarding θ, it was interpreted by Laplace that Bayes used a uniform prior in his "Scholium". An objection raised by Fisher to this is essentially on the grounds of a lack of invariance. He argued that setting a parameter θ to be uniform resulted in, say $\tau = \theta^3$ (or $\tau = \tau(\theta)$) and then why not set τ to be uniform so that $g(\tau) = 1, \ 0 < \tau < 1$ then implies that $g(\theta) = 3\theta^2$ instead $g(\theta) = 1$. Hence one will get different answers depending on what function of the parameter is assumed uniform.

Jeffreys countered this lack of invariance with the following:
The Fisher Information quantity of a probability function $f(x|\theta)$ is

$$I(\theta) = E\left(\frac{d \log f}{d\theta}\right)^2,$$

assuming it exists. Then set

$$g(\theta) = I^{\frac{1}{2}}(\theta).$$

Now suppose $\tau = \tau(\theta)$. Then

$$I(\tau) = E\left(\frac{d \log f}{d\tau}\right)^2 = E\left(\frac{d \log f}{d\theta} \times \frac{d\theta}{d\tau}\right)^2$$

$$= E\left(\frac{d \log f}{d\theta}\right)^2 \times \left(\frac{d\theta}{d\tau}\right)^2,$$

or

$$I(\tau)\left(\frac{d\tau}{d\theta}\right)^2 = I(\theta),$$

and thus

$$I^{\frac{1}{2}}(\tau)d\tau = I^{\frac{1}{2}}(\theta)d\theta.$$

So if you start with the prior for θ to be $g(\theta) \propto I^{\frac{1}{2}}(\theta)$ this leads to $g(\tau) \propto I^{\frac{1}{2}}(\tau)$.
Note also if we set

$$\tau = \int_{-\infty}^{\theta} I^{\frac{1}{2}}(\theta')d\theta'$$

then

$$d\tau = I^{\frac{1}{2}}(\theta)d\theta$$

and τ is uniform. At any rate, Jeffreys solved the invariance problem for the binomial by noting that

$$I(\theta) = E\left(\frac{d\log f}{d\theta}\right)^2 = -E\left(\frac{d^2\log f}{d\theta^2}\right)$$

and consequently, for the Binomial

$$\log f = \log\binom{n}{r} + r\log\theta + (n-r)\log(1-\theta)$$

$$\frac{d\log f}{d\theta} = \frac{r}{\theta} - \frac{(n-r)}{1-\theta}, \qquad \frac{d^2\log f}{d\theta^2} = -\frac{r}{\theta^2} - \frac{(n-r)}{(1-\theta)^2}$$

$$-E\left(\frac{d^2\log f}{d\theta^2}\right) = \frac{n\theta}{\theta^2} + \frac{(n-n\theta)}{(1-\theta)^2} = n\left(\frac{1}{\theta} + \frac{1}{1-\theta}\right) = \frac{n}{\theta(1-\theta)},$$

and consequently

$$I_n^{\frac{1}{2}}(\theta) \propto \frac{1}{\theta^{\frac{1}{2}}(1-\theta)^{\frac{1}{2}}}. \tag{6.3.1}$$

So

$$g(\theta) \propto \frac{1}{\theta^{\frac{1}{2}}(1-\theta)^{\frac{1}{2}}},$$

$$p(\theta|r) \propto \theta^{r-\frac{1}{2}}(1-\theta)^{n-r-\frac{1}{2}},$$

while use of the uniform prior yields $p(\theta|r) \propto \theta^r(1-\theta)^{n-r}$.

Example 6.5

Suppose

$$f_r(n|r) = \binom{n-1}{r-1} \theta^r (1-\theta)^{n-r}, \quad n = r, r+1, \ldots$$

or

$$f_{n-r}(n|n-r) = \binom{n-1}{n-r-1} \theta^r (1-\theta)^{n-r}$$

$$= \binom{n-1}{r} \theta^r (1-\theta)^{n-r}, \quad n = r+1, r+2, \ldots$$

Now by calculating

$$I_r^{\frac{1}{2}}(\theta) \propto \theta^{-1}(1-\theta)^{-\frac{1}{2}} \tag{6.3.2}$$

$$I_{n-r}^{\frac{1}{2}}(\theta) \propto \theta^{-\frac{1}{2}}(1-\theta)^{-1}, \tag{6.3.3}$$

we have demonstrated that three different Jeffreys' priors result in three different posteriors for θ, thus, in a sense, contravening the likelihood principle.

Now consider the problem of predicting the number of successes T out of a total M when we have seen r successes and $n - r$ failures. Note M includes the original n trials and T includes the original r successes starting with the priors (6.3.1), (6.3.2) and (6.3.3) and that these predictive probabilities for T differ for the three priors although their likelihoods are the same.

Now suppose we have an urn with a known number M of balls marked s and f of which an unknown number t are marked s. The object is to infer the unknown number t after taking a sample from the urn that yields r successes and $n - r$ failures in any one of the following three ways:

1. Sample $n < M$. Then we obtain the hypergeometric

$$p(r|n, M, t) = \binom{t}{r}\binom{M-t}{n-r} \Big/ \binom{M}{n}$$

$r = 0, 1, \ldots, \min(t, n)$.

2. Sampling until r successes yields

$$p(n|r, M, t) = \binom{n-1}{r-1}\binom{M-n}{t-r} \Big/ \binom{M}{t}$$

for $n = r, \ldots, \min(M, r + M - t)$;

3. Sampling until $n - r$ failures yields

$$p(n|n-r, M, t) = \binom{n-1}{n-r-1}\binom{M-n}{(M-t)-(n-r)} \bigg/ \binom{M}{M-t},$$

for $n = r, r+1, \ldots, \min(M, r+M-t)$. Note the likelihood $L(t)$ is the same for each of the above.

$$L(t) = \frac{t!(M-t)!}{(t-r)![M+r-t-n]!}.$$

When little is known about a finite number of discrete possibilities for t it appears sensible to assume equal probabilities for each possibility or

$$p(t|M) = \frac{1}{(M+1)} \quad t = 0, 1, \ldots, M,$$

so

$$p(t|r, n, M) = \frac{\binom{t}{r}\binom{M-t}{n-r}}{\binom{M+1}{n+1}} \quad t = r, r+1, \ldots, \min(M-n+r, n).$$

Now for large M

$$p(r|n, M, t) \doteq \binom{n}{r}\left(\frac{t}{M}\right)^r\left(1-\frac{t}{M}\right)^{n-r},$$

$$p(n|r, M, t) \doteq \binom{n-1}{r-1}\left(\frac{t}{M}\right)^r\left(1-\frac{t}{M}\right)^{n-r},$$

$$p(n|n-r, M, t) \doteq \binom{n-1}{n-r-1}\left(\frac{t}{M}\right)^r\left(1-\frac{t}{M}\right)^{n-r},$$

and

$$P\left[\frac{t}{M} \le z|M, n, r\right] \doteq \frac{\Gamma(n+2)}{\Gamma(r+1)\Gamma(n-r+1)}\int_0^z x^r(1-x)^{n-r}dx. \tag{6.3.4}$$

Under appropriate conditions as M grows $tM^{-1} \to \theta$ and the left sides of the above tend to the right sides and the right side of (6.3.4) is the posterior distribution function of θ for the uniform prior $g(\theta) = 1$.

Keeping the problem discrete and finite avoids the criticism that the uniformity of $\frac{1}{M}$, which approaches, θ does not imply the uniformity of any monotonic function of $\frac{1}{M} \to \theta$ which Fisher leveled at Laplace. It also conforms to the Likelihood Principle.

In much scientific, technical, and medical experimentation in which individual trials are binary, the parametric Bernoulli model is used as a basis for the analysis of the data, assuming independent copies with its resulting likelihood function. Although this is a useful and convenient paradigm for this type of trial, it is not as appropriate in most instances as one stipulating that only a finite number of trials can be made, no matter how large the number. The first model assumes that there is some value θ that is the probability of success on each individual trial, and the second entertains the notion that from a finite number of binary events, either having occurred or having the potential to occur, a certain number of successes are observed that are in no way distinguishable from the rest before being observed. The latter model can actually include the first even when the first has some legitimate claim to describing the process, such as repeated tossing of the same coin. We assume the tosses are generated so that there is a sequence of heads and tails and that some fraction of the sequence is then observed. One of the advantages of the second approach for a Bayesian is the greater simplicity (difficult though it is) of thinking about prior distributions of the observable quantities—say the number of heads out of the total—over trying to focus on a distribution for a hypothetical parameter.

The finite model that focuses on the fraction of successes in the finite population is basically a dependent Bernoulli model that leads to the hypergeometric likelihood, common to the particular cases

$$L(t) = \frac{t!(M-t)!}{(t-r)!(M-t-n+r)!}.$$

In many situations we are interested in the chance that the next observation is a success. This can be calculated by letting $M = n+1$ and is useful in determining the chance that a therapy already given to n ailing people more or less similar to a patient, and having cured t of them, will also cure that patient of the ailment. A physician who wishes to know the chance that a particular fraction of a given number of patients, say $M - n$, whom he is treating will be cured can calculate

$$P[(t-r)/(M-n) \le z|M, n, r].$$

A pharmaceutical company or a government health organization may assume that the number of potential future cases is sufficiently large that an asymptotic approximation is accurate enough to provide adequate information regarding the cured fraction of that finite, though not necessarily specified, number of cases. In cases in which only some normative evaluation is required, the probability function for success on the next observation and for the fraction of successes as M grows would be informative. The connection between the finite model and the existence of the parameter θ is given by the representation theorem of de Finetti which entails an infinite sequence of exchangeable random variables.

Theorem 6.1   To every infinite sequence of exchangeable random binary variables $\{W_k\}$ corresponds a probability distribution concentrated on [0,1]. For every permutation of the integers $1, \ldots, M$, and for any given integer $r \in [0, M]$, let P_r represent probability defined on M-variate sequences of r ones and $M - r$ zeros. Then

(a) $P_r[W_{j_1} = 1, \ldots, W_{j_r} = 1, W_{j_{r+1}} = 0, \ldots, W_{j_M} = 0] \equiv P_{rM} = \int_0^1 \theta^r (1 - \theta)^{M-r} dF(\theta),$

(b) $P_r[\sum_{k=1}^M W_k = r] = \binom{M}{r} \int_0^1 \theta^r (1 - \theta)^{M-r} dF(\theta),$

(c) $\lim_{M \to \infty} M^{-1} \sum_{k=1}^M W_k = \theta,$

(with probability one) with θ having distribution function $F(\theta)$.

   Heath and Sudderth (1976) devised the following simple proof.

Lemma 6.1   Suppose $W_1, \ldots, W_m$ is a sequence of exchangeable binary random variables and let

$$q_t = P_r\left(\sum_{j=1}^m W_j = t\right).$$

Then for $0 \le r \le M \le m$

$$P_{rM} = \sum_{t=r}^{r+m-M} \frac{(t)_r (m - t)_{M-r}}{(m)_M} q_t,$$

where $(x)_r = \prod_{j=0}^{r-1} (x - j).$

Proof: It follows from exchangeability that, given $\sum_{j=1}^m W_j = t$, all $m!$ possible permutations of the m distinct zeros and ones are equally likely so that the probability that there are exactly r ones followed by exactly $M - r$ zeros followed by any permutation of $t - r$ ones and $m - M - (t - r)$ zeros is

$$\frac{(t)_r (m - t)_{M-r} (m - M)!}{m!} = \frac{(t)_r (m - t)_{M-r}}{(m)_M}.$$

The result follows from recognizing that the event {first r components are one and the next $M - r$ are zero} $= \bigcup_{t=r}^{r+m-M}$ {first r components are one, the next $M - r$ are zero, $t - r$ out of the next $m - M$ are one} which justifies P_{rM}.   □

   Now apply the lemma to $W_1, \ldots, W_m$ so that

$$P_{rM} = \int_0^1 \frac{(\theta m)_r ((1 - \theta)m)_{M-r}}{(m)_M} F_m(d\theta),$$

where F_m is the distribution function concentrated on $\{\frac{t}{m} : 0 \le t \le m\}$ whose jump at $\frac{t}{m}$ is q_t. Then apply Helly's theorem *cf.* Feller (1966) to get a subsequence that converges in distribution to a limit F. Since m tends to infinity the integrand above tends uniformly to the integrand in (a) so (a) holds for this F.

The theorem can be extended to show that every infinite sequence of exchangeable variables is a mixture of independent, identically distributed random variables, *cf.* Loeve (1960).

The importance of the representation theorem is that it induces a parameter which has been lurking in the background of the finite case as the limit of a function of observables. Thus it enables consideration of a parametric model in cases well beyond the simple error model

$$X = \theta + e,$$

where θ was some physical entity or constant.

6.4 COHERENCE

Example 6.6

A situation termed Dutch Book, or a "Day at the Races," can be described as follows: Bettor A, who has S dollars to bet on a 3-horse race is offered the following odds by Bookmaker B:

	Horse 1 2/1	Horse 2 3/1	Horse 3 4/1	
Winning Horse	Some Potential Outcomes			
	1	2	3	
A Bets	S	0	0	
B's Gain	$-2S$	S	S	
A Bets	0	S	0	
B's Gain	S	$-3S$	S	
A Bets	0	0	S	
B's Gain	S	S	$-4S$	
A Bets	$S/3$	$S/3$	$S/3$	
B's Gain	0	$-S/3$	$-2S/3$	B cannot win
A Bets	$(4/9)S$	$(3/9)S$	$(2/9)S$	
B's Gain	$-S/3$	$-S/3$	$-S/9$	B is a sure loser
A Bets	$\frac{20}{47}S$	$\frac{15}{47}S$	$\frac{12}{47}S$	
B's Gain	$\frac{-13}{47}S$	$\frac{-13}{47}S$	$\frac{-13}{47}S$	B loses the same amount no matter which horse wins.

According to B, the implicit probability of horses 1, 2 and 3 winning are $1/3$, $1/4$, and $1/5$ respectively. In general with n horses suppose B offers odds $\frac{1-P_j}{P_j}$ on Horse j to win so if A's total wager is $\sum_{j=1}^{n} S_j P_j$ then B's gain if Horse t wins is

$$G_t = \sum_{j=1}^{n} S_j P_j - S_t P_t - (1 - P_t)S_t = \sum_{j=1}^{n} S_j P_j - S_t.$$

Now if $\sum P_j < 1$ and letting $S_j = S/n$, then

$$\sum S_j P_j = \frac{S}{n} \sum P_j,$$

and

$$G_t = \frac{S}{n}\left(\sum_j P_j - 1\right)$$

represents a loss to B no matter which horse wins. Thus under these circumstances, it would seem that the bookmaker should give odds that cohere with the laws of probability.

Suppose we have a discrete probability function

$$P(D_i|\theta_j) = p_{ij} > 0 \quad i = 1, \ldots, n \text{ and } j = 1, \ldots, N.$$

A Master of Ceremonies chooses a θ_j, does the chance experiment, and announces the realized D_i and p_{ij}. Denote every distinct subset of values of $\theta_1, \ldots, \theta_N$ by I_k where $k = 1, 2, \ldots, 2^N$. Now the problem of statistician B is to make probability assignments P_{ik} for all 2^N subsets knowing the value p_{ij} and the sampled value D_i that was realized. When θ_j is true, the subset is considered correct if it includes θ_j and incorrect if it doesn't. An antagonist A may bet for or against any combination of the 2^N subsets also knowing p_{ij} and the realized D_i. Thus B assigns P_{ik} and A assigns a stake S_{ik} for the subset I_k. A then gives amount $P_{ik}S_{ik}$ to B and receives in turn S_{ik} if I_k is correct and 0 if it is incorrect. A is risking $P_{ik}S_{ik}$ to win $(1 - P_{ik})S_{ik}$ (i.e. by being offered odds of $1 - P_{ik}$ to P_{ik} on I_k if $S_{ik} > 0$ and P_{ik} to $1 - P_{ik}$ if $S_{ik} < 0$). When $\theta = \theta_j$, B's gain on this bet when D_i is realized is

$$G_{ijk} = (P_{ik} - \delta_{jk})S_{ik},$$

where

$$\delta_{jk} = 1 \text{ if } \theta_j \in I_k \quad \text{so } G_{ijk} = (P_{ik} - 1)S_{ik}$$
$$\delta_{jk} = 0 \text{ if } \theta_j \notin I_k \quad \text{so } G_{ijk} = P_{ik}S_{ik}.$$

Now

$$G_{ij} = \sum G_{ijk} = \sum_k (P_{ik} - \delta_{jk})S_{ik}$$

and the expected gain when $\theta = \theta_j$ is

$$G_{.j.} = \sum_i p_{ij} \sum_k (P_{ik} - \delta_{jk})S_{ik}.$$

Now for any $\{S_{ik}\}$, if $G_{.j.} \leq 0$ for all j and $G_{.j.} < 0$ for at least one value of j we say B's probability assignment is incoherent since on the average he will lose money for at least one θ_j or more and at best holding his own for the others. Otherwise we say that his probability assignments are coherent. Note if a set of finite positive q_j's exist such that

$$\sum_{j=1}^{N} q_j G_{.j.} = 0,$$

then the probability assignment is coherent since either $G_{.j.} = 0$ for all j or if $G_{.j.} < 0$ for some values of j there must be at least one value of j such that $G_{.j.} > 0$.

Theorem 6.2 A Bayesian assignment of probabilities by B is both necessary and sufficient for coherence. (Given a Bayesian assignment we have coherence and given coherence the assignment must be Bayesian.) The result is basically due to de Finetti (1937).

The following proof is due to Cornfield (1969):

Proof: **Sufficiency** A Bayesian assignment for the prior probability for θ_j can be proportional to values $q_j > 0$. In other words the posterior probability is

$$P(\theta = \theta_j | D = D_i) = \frac{q_j p_{ij}}{\sum_{j=1}^{N} q_j p_{ij}},$$

which implies that

$$P(\theta \in I_k | D_i) = \sum_{j=1}^{N} q_j p_{ij} \delta_{jk} \Big/ \sum_{j=1}^{N} q_j p_{ij}. \tag{6.4.1}$$

Then define $P_{ik} = P(\theta \in I_k | D_i)$. In this setup the likelihood is given by p_{ij} and the prior probabilities given by

$$q'_j = q_j \Big/ \sum_{t=1}^{N} q_t.$$

The above definition of P_{ik} (6.4.1) implies that

$$\sum_j p_{ij} q_j (P_{ik} - \delta_{jk}) = 0.$$

Therefore, for all $\{S_{ik}\}$

$$\sum_i \sum_k S_{ik} \sum_j p_{ij} q_j (P_{ik} - \delta_{jk}) = 0$$

or, by interchanging summation signs,

$$0 = \sum_j q_j \sum_i p_{ij} \sum_k (P_{ik} - \delta_{jk}) S_{ik} = \sum_j q_j G_{\cdot j}.$$

Therefore, the probability assignment is coherent and the sufficiency is established.
□

Necessity: To establish necessity, we need to show that P_{ij} must satisfy the axioms of probability:

1. $0 \leq P_{ij} \leq 1$ for all i and j.
2. If $I_k = I_{k_1} \cup I_{k_2}$ and $I_{k_1} \cap I_{k_2} = \emptyset$ then $P_{ik} = P_{ik_1} + P_{ik_2}$ for all i, k_1 and k_2
3. $\sum_j P_{ij} = 1$ for all i

To establish (1) let A select S_{ij}. Then B's gain is $P_{ij} S_{ij} - \delta_{ij}$ if θ_j is true and $P_{ij} S_{ij}$ when θ_j is not true. If $P_{ij} < 0$, A sets $S_{ij} > 0$ and B loses no matter whether θ_j is true or not. Similarly if $P_{ij} > 1$, A sets $S_{ij} < 0$ and again B loses. Hence given coherence (1) holds.

To establish (3), let A choose S_i for each θ_j, $j = 1, \ldots, N$ thus risking $S_i \sum_j P_{ij}$. Since one of the θ_j's must obtain, A is paid S_i, so B's gain is $S_i \left(\sum_j P_{ij} - 1 \right)$, which can be made negative by the sign of S_i opposite the sign of $\sum_j P_{ij} - 1$, unless it is zero.

To prove (2), let A select S_{ij_1}, S_{ij_2}, and S_{ik}. Then B's gain for D_i when θ_{j_1} obtains, θ_{j_2} obtains and when neither obtains are

$$\begin{aligned}
G_{ij_1} &= (P_{ij_1} - 1)S_{ij_1} + P_{ij_2} S_{ij_2} + (P_{ik} - 1)S_{ik}, \\
G_{ij_2} &= P_{ij_1} S_{ij_1} + (P_{ij_2} - 1)S_{ij_2} + (P_{ik} - 1)S_{ik}, \qquad (6.4.2) \\
G_{ij_3} &= P_{ij_1} S_{ij_1} + P_{ij_2} S_{ij_2} + P_{ik} S_{ik}.
\end{aligned}$$

Thus A must find solutions to the above equations equal to specified negative values.

To prevent this B must choose P_{ij} such that

$$\begin{vmatrix} P_{ij_1} - 1 & P_{ij_2} & P_{ik} - 1 \\ P_{ij_1} & P_{ij_2} - 1 & P_{ik} - 1 \\ P_{ij_1} & P_{ij_2} & P_{ik} \end{vmatrix} = 0 \qquad (6.4.3)$$

or $P_{ik} = P_{ij_1} + P_{ij_2}$.

Next we show that $P_{ij} \propto p_{ij}$. It is sufficient to consider any two D_1 and D_2 say and sets I_1 and I_1^C. Now B has probability assignment P_{ij} to θ_j and $\sum_{j_1} P_{ij_1} \delta_{j_1 1}$ to I_1.
A selects the following:

	I_1	I_1^C
D_1	$S_{11} = kS$	$S_{12} = 0$
D_2	$S_{21} = -k$	$S_{22} = 0$
$D_3, \ldots, D_N$	$S_{i1} = 0$	$S_{i2} = 0.$

Now B's expected gain is

$$G_1 = p_{1j_1} kS \left(\sum_j P_{1j} \delta_{j1} - 1 \right) - p_{2j_1} k \left(\sum_j P_{2j} \delta_{j1} - 1 \right) \qquad \text{for all } \theta_{j_1} \in I_1.$$

The expected gain is

$$G_2 = p_{1j_2} kS \sum_j P_{1j} \delta_{j1} - p_{2j_2} k \sum_j P_{2j} \delta_{j1} \qquad \text{for all } \theta_{j_2} \in I_1^C.$$

Now when

$$\frac{p_{2j_1}}{p_{1j_1}} \times \frac{(1 - \sum_j P_{2j} \delta_{j1})}{(1 - \sum_j P_{1j} \delta_{j1})} < S < \frac{p_{2j_2}}{p_{1j_2}} \times \frac{\sum_j P_{2j} \delta_{j1}}{\sum_j P_{1j} \delta_{j1}} \qquad (6.4.4)$$

for all $\theta_{j_1} \in I_1, \theta_{j_2} \in I_2$ and $k > 0$ or

$$\frac{p_{2j_1}}{p_{1j_1}} \times \frac{(1 - \sum_j P_{2j} \delta_{j1})}{(1 - \sum_j P_{1j} \delta_{j1})} > S > \frac{p_{2j_2}}{p_{1j_2}} \times \frac{\sum_j P_{2j} \delta_{j1}}{\sum_j P_{j1} \delta_{j1}}$$

for all $\theta_{j_1} \in I_1, \theta_{j_2} \in I_1^C$ and $k < 0$, G_1 and G_2 will be negative.

By selection of k, A can induce losses whether or not $\theta_j \in I_1$ or I_1^C unless B chooses P_{ij} so that for every partition of $(\theta_1, \ldots, \theta_N)$ into two sets,

$$\frac{\sum_j P_{2j} \delta_{j2}}{\sum_j P_{1j} \delta_{j2}} \div \frac{\sum_j P_{2j} \delta_{j1}}{\sum_j P_{1j} \delta_{j1}} = \frac{p_{2j_2}}{p_{1j_2}} \div \frac{p_{2j_1}}{p_{1j_1}}$$

for $\theta_{j_1} \in I_1, \theta_{j_2} \in I_1^C = I_2$. This requires that

$$\frac{P_{2j_2}}{P_{1j_2}} \div \frac{P_{2j_1}}{P_{1j_1}} = \frac{p_{2j_2}}{p_{1j_2}} \div \frac{p_{2j_1}}{p_{1j_1}} \tag{6.4.5}$$

for all θ_{j_1} and θ_{j_2}, since if the above is not satisfied A can select a set I_1 for which the inequalities of (6.4.4) hold. Equation (6.4.5) implies that

$$P_{ij} \propto p_{ij}q_j \quad \text{for} \quad i = 1, 2 \text{ and all } j$$

because of (3). This applies for every pair of sample values so that the above obtain for all i. Now the q_i must all be of the same sign otherwise (1) will not hold. It is also necessary that $q_i \neq 0$ since if the j_1th is, A sets $S_{ij} = 0$ for all θ_j except $\theta_j = \theta_{j_1}$ and wins S_{ij_1} when θ_{j_1} obtains by risking $P_{ij_1}S_{ij_1}$ that is, zero, making B's assignment incoherent. Necessity now has been established.

What the result implies is that if B's betting odds are not consistent with prior probabilities then A by appropriate choice of stakes can render B a sure loser. So only the Bayes approach can claim a universal consistency or coherence in the jargon of betting. Whether betting is always an appropriate metaphor for scientific inference is still a debateable issue.

For a more general result see Freedman and Purves (1969).

6.5 MODEL SELECTION

A more general Bayesian approach that includes Hypothesis Testing goes under the rubric of Model Selection. Suppose there are K possible models $M_1, \ldots, M_K$ that could explain the generation of a data set $X^{(N)} = x^{(N)}$. Let

$$P(M_k) = q_k, \quad k = 1, \ldots, K, \quad \sum_k q_k = 1$$

and associated with M_k we have probability function

$$f(x^{(N)}, \theta_k | M_k) = f(x^{(N)} | \theta_k, M_k)p(\theta_k | M_k),$$

where $p(\theta_k | M_k)$ specifies a prior probability function for θ_k under model M_k. Now

$$f(x^{(N)} | M_k) = \int f(x^{(N)}, \theta_k | M_k)d\theta_k.$$

Further,

$$P(M_k | x^{(N)}) = \frac{q_k f(x^{(N)} | M_k)}{\sum_j q_j f(x^{(N)} | M_j)} = q_k'.$$

Selection of the model could depend on a loss function. If the loss function assigns the same loss to each of the entertained models then one would choose that model M_{k^*} associated with

$$q'_{k^*} = \max_k q'_k,$$

that is, the model with the largest posterior probability.

Almost all of Bayesian Hypothesis Testing problems can be subsumed under the Model Selection paradigm. Note that Examples 6.1–6.4 all can be considered as Model Selection problems.

Example 6.7

Suppose we have a set of data with two labels such that the realized values of

$$X^{(N)} = x^{(N)} = (x^{(N_1)}, x^{(N_2)}), N_1 + N_2 = N$$

$$X^{(N_1)} = x^{(N_1)} = (x_1, \ldots, x_{N_1}), \quad X^{(N_2)} = x^{(N_2)} = (x_{N_1+1}, \ldots, x_{N_1+N_2}).$$

Assume the model specifications are

$$M_1 : X_i \quad i = 1, \ldots, N_1 + N_2 \text{ are i.i.d. with density}$$
$$f(x|\theta) = \theta e^{-\theta x}$$
$$p(\theta|M_1) \propto \theta^{\delta-1} e^{-\gamma\theta},$$

with prior probability q_1 and

$$M_2 : X_i, \ i = 1, \ldots, N_1 \text{ are i.i.d. with density}$$
$$f(x|\theta) = \theta_1 e^{-\theta_1 x}$$
$$p(\theta_1|\delta_1, \ \gamma_1) \propto \theta_1^{\delta_1-1} e^{-\gamma_1 \theta_1}$$

independent of

$$X_i, \ i = N_1 + 1, \ldots, N_1 + N_2 \text{ that are i.i.d. with density}$$
$$f(x|\theta) = \theta_2 e^{-\theta_2 x}$$
$$p(\theta_2|\gamma_2, \ \delta_2) \propto \theta_2^{\delta_2-1} e^{-\gamma_2 \theta_2}$$

with prior probability $q_2 = 1 - q_1$.

Assuming that δ, γ, δ_1, γ_1, δ_2, γ_2 and q_1 are known and losses for selection are equal, then choice of the best model depends on the larger of

$$q_1 f(x^{(N)}|M_1) = q_1 \Gamma(N+\delta)\gamma^\delta \Big/ \Gamma(\delta)[N\bar{x} + \gamma]^{N+\delta} \propto q'_1$$

or

$$q_2 \, f(x^{(N)}|M_2) = q_2 f(x^{(N_1)}|M_2) f(x^{(N_2)}|M_2) \propto q_2',$$

where

$$f(x^{(N_j)}|M_2) = \Gamma(N_j + \delta_j)\gamma_j^{\delta_j} \Big/ \Gamma(\delta_j)[N_j\bar{x}_j + \gamma_j]^{N_j+\delta_j}.$$

REFERENCES

Cornfield, J. (1969). The Bayesian outlook and its applications. *Biometrics*, **25, 4**, 617–657.

de Finetti, B. (1973). Le Prevision: ses lois logiques, ses sources subjectives. *Annals Institute Poincaire*, ***VIII***, fasc. 1, 1–68. Reprinted in Studies in Subjective Probability. Melbourne, FL: Krieger, 1980 (English translation).

Freedman, D. A. and Purves, R. A. (1969). Bayes' method for bookies. *Annals of Mathematical Statistics*, **40**, 1177–1186.

Heath, D. and Sudderth, W. D. (1976). de Finetti's theorem on exchangeable variables. *American Statistician*, **7** (4), 718–728.

Jeffreys, H. (1961). *Theory of Probability*. Clarendon Press.

Loeve, M. (1960). *Probability Theory*. New York: Van Nostrand.

CHAPTER SEVEN

Theories of Estimation

This chapter presents a thorough investigation of frequentist and likelihood based point estimation. The concepts of unbiasedness, consistency and Fisher consistency are introduced and illustrated. Error bounds on estimation are presented including the classic Cramér-Rao lower bound and its generalizations. Fisher information is interpreted from several points of view, and its relation to efficiency in estimation is laid out. The Blackwell-Rao theorem is established. Then considerable attention is paid to large sample theory for maximum likelihood estimation; asymptotic consistency and normality are established. The chapter concludes with definitions and some discussion of sufficiency principles.

7.1 ELEMENTS OF POINT ESTIMATION

Essentially there are three stages of sophistication with regard to estimation of a parameter:

1. At the lowest level—a simple point estimate;
2. At a higher level—a point estimate along with some indication of the error of that estimate;
3. At the highest level—one conceives of estimating in terms of a "distribution" or probability of some sort of the potential values that can occur.

This entails the specification of some set of values presumably more restrictive than the entire set of values that the parameter can take on or relative plausibilities of those values or an interval or region.

Consider the I.Q. of University of Minnesota freshmen by taking a random sample of them. We could be satisfied with the sample average as reflective of that population. More insight, however, may be gained by considering the variability

Modes of Parametric Statistical Inference, by Seymour Geisser
Copyright © 2006 John Wiley & Sons, Inc.

of scores by estimating a variance. Finally one might decide that a highly likely interval for the entire average of freshmen would be more informative.

Sometimes a point estimate is about all you can do. Representing distances on a map, for example. At present there is really no way of reliably illustrating a standard error on a map—so a point estimate will suffice.

An "estimate" is a more or less reasonable guess at the true value of a magnitude or parameter or even a potential observation, and we are not necessarily interested in the consequences of estimation. We may only be concerned in what we should believe a true value to be rather than what action or what the consequences are of this belief. At this point we separate estimation theory from decision theory, though in many instances this is not the case.

Example 7.1

Suppose I am building a fence and I estimate I need about 200 linear feet of lumber. I may only order 180 feet so as not to incur wastage (in money) or I may order 220 feet so as not to have to make two orders. While 220 feet and 180 feet may end up being my decisions they are not my estimates, though they may modify the estimate to conform with a loss criterion.

In other words decision making should not be confused with estimating or guessing.

7.2 POINT ESTIMATION

1. An estimate might be considered "good" if it is in fact close to the true value on average or in the long run (pre-trial).
2. An estimate might be considered "good" if the data give good reason to believe the estimate will be close to the true value (post trial).

A system of estimation will be called an estimator

1. Choose estimators which on average or very often yield estimates which are close to the true value.
2. Choose an estimator for which the data give good reason to believe it will be close to the true value that is, a well-supported estimate (one that is suitable after the trials are made.)

With regard to the first type of estimators we do not reject one (theoretically) if it gives a poor result (differs greatly from the true value) in a particular case (though you would be foolish not to). We would only reject an estimation procedure if it gives bad results on average or in the long run. The merit of an estimator is judged, in general, by the distribution of estimates it gives rise to—the properties of its sampling distribution. One property sometimes stressed is unbiasedness. If

$T(D)$ is the estimator of θ then unbiasedness requires

$$E[T(D)] = \theta.$$

For example an unbiased estimator of a population variance σ^2 is

$$(n-1)^{-1} \sum (x_i - \bar{x})^2 = s^2$$

since $E(s^2) = \sigma^2$.

Suppose $Y_1, Y_2, \ldots$ are i.i.d. Bernoulli random variables $P(Y_i = 1) = \theta$ and we sample until the first "one" comes up so that probability that the first one appears after $X = x$ zeroes is

$$P(X = x|\theta) = \theta(1-\theta)^x \quad x = 0, 1, \ldots \quad 0 < \theta < 1.$$

Seeking an unbiased estimator we have

$$\theta = E(T(Y)) = \sum_{x=0}^{\infty} t(x)\theta(1-\theta)^x = t(0)\theta + t(1)\theta(1-\theta) + \cdots.$$

Equating the terms yields the unique solution $t(0) = 1$, $t(x) = 0$ for $x \geq 1$. This is flawed because this unique estimator always lies outside of the range of θ. So unbiasedness alone can be a very poor guide. Prior to unbiasedness we should have consistency (which is an asymptotic type of unbiasedness, but considerably more). Another desideratum that many prefer is invariance of the estimation procedure. But if $E(X) = \theta$, then for $g(X)$ a smooth function of X, $E(g(X)) \neq g(\theta)$ unless $g(\cdot)$ is linear in X. Definitions of classical and Fisher consistency follow:

Consistency: An estimator T_n computed from a sample of size n is said to be a consistent estimator of θ if for any arbitrary $\epsilon > 0$ and $\delta > 0$ there is some value, N, such that

$$P[|T_n - \theta| < \epsilon] > 1 - \delta \quad \text{for all } n > N,$$

so that T_n converges in probability to θ. This is a limiting property and doesn't say how to produce an estimator. Note this permits, for fixed a and b, that $(n-a)/(n-b)T_n$ to be consistent if T_n is consistent. Similarly $T_n + k_n$ will then also be consistent for any $k_n \to 0$. Let T' stand for any arbitrary function of the first n_1 observations and T''_{n-n_1} be such that for all $n > n_1$, $T'' \to \theta$ as n increases. Then if $n > n_1$ let $T_n = \frac{1}{n}[n_1 T' + (n - n_1)T''_{n-n_1}]$ and if $n \leq n_1$, $T_n = T'$, so that T_n is defined for all values of n and is consistent and foolish for $n \leq n_1$. Clearly it is perfectly arbitrary and potentially perfectly useless as Fisher (1956) pointed out.

Fisher's Definition of Consistency for i.i.d. Random Variables

"A function of the observed frequencies which takes on the exact parametric value when for those frequencies their expectations are substituted."

For a discrete random variable with $P(X_j = x_j|\theta) = p_j(\theta)$ let T_n be a function of the observed frequencies n_j whose expectations are $E(n_j) = np_j(\theta)$. Then the linear function of the frequencies $T_n = \frac{1}{n}\sum_j c_j n_j$ will assume the value

$$\tau(\theta) = \Sigma c_j p_j(\theta),$$

when $np_j(\theta)$ is substituted for n_j and thus $n^{-1}T_n$ is a consistent estimator of $\tau(\theta)$.

Another way of looking at this is:

$$\text{Let } F_n(x) = \frac{1}{n} \times \# \text{ of observations } \le x$$

$$= \frac{i}{n} \text{ for } x_{(i-1)} < x \le x_{(i)},$$

where $x_{(j)}$ is the jth smallest observation. If $T_n = g(F_n(x))$ and $g(F(x|\theta)) = \tau(\theta)$ then T_n is Fisher consistent for $\tau(\theta)$. Note if

$$T_n = \int xdF_n(x) = \bar{x}_n,$$

and if

$$g(F) = \int xdF(x) = \mu,$$

then $\bar{x}_n$ is Fisher consistent for μ.

On the other hand, if $T_n = \bar{x}_n + \frac{1}{n}$, then this is not Fisher consistent but is consistent in the ordinary sense. Fisher Consistency is only defined for i.i.d. $X_1, \ldots, X_n$.

However, as noted by Barnard (1974), "Fisher consistency can only with difficulty be invoked to justify specific procedures with finite samples" and also "fails because not all reasonable estimates are functions of relative frequencies." He also presents an estimating procedure that does meet his requirements that the estimate lies within the parameter space and is invariant based on pivotal functions.

Example 7.2

For $X_1, \ldots, X_n$ i.i.d. $N(\mu, \sigma^2)$, let

$$t = \frac{\sqrt{n}(\bar{x} - \mu)}{s},$$

whose distribution does not depend on the parameters μ and σ. Barnard points out that setting t equal to its mean (or median), for example,

$$\frac{\sqrt{n}(\bar{x} - \mu)}{s} = 0$$

results in the estimate $\hat{\mu} = \bar{x}$. If $\tau = g(\mu)$ is a smooth function of μ such that

$$\frac{\sqrt{n}(\bar{x} - g^{-1}(\tau))}{s} = 0,$$

then

$$g(\bar{x}) = \hat{\tau}.$$

A well-supported estimator could be defined thus: If for all $a > 0$ and given data D,

$$\text{Support } [|T_n - \theta| < a|D] \geq \text{Support } [|T_n^* - \theta| < a|D],$$

then T_n is better supported than T_n^*. Support can be the likelihood or some other measure of plausibility given D—a post data evaluation. We can also define an interval for θ such that $L(\theta) \leq KL(\theta)$, where $K > 1$ for all θ

$$\text{or} \quad \frac{L(\theta)}{L(\hat{\theta})} \geq \frac{1}{K}.$$

In comparing estimators in frequentist terms, one criterion for T_n to be superior to T_n^* could be

$$P\{|T_n - \theta| < a|\theta\} \geq P\{|T_n^* - \theta| < a|\theta\},$$

for all $a > 0$, and all n. Then we would say that T_n is closer to θ than T_n^*. If this is true for all competitors T_n^*, then we could say that T_n is the uniformly best estimator. Unfortunately such estimators seem impossible to find for every n, and all values of θ. If we restrict ourselves to the class of all estimators that are asymptotically normal and unbiased then for a large enough n, what is required is that T_n have the smallest variance since if

$$T \sim N(\theta, \sigma^2) \quad \text{and} \quad T^* \sim N(\theta, \tau^2) \quad \text{and} \quad \sigma^2 \leq \tau^2,$$

then it is easy to show that for all $a > 0$

$$P(|T - \theta| \le a) = \Phi\left(\frac{a}{\sigma}\right) - \Phi\left(-\frac{a}{\sigma}\right),$$

$$P(|T^* - \theta| \le a) = \Phi\left(\frac{a}{\tau}\right) - \Phi\left(-\frac{a}{\tau}\right).$$

Since $\frac{a}{\sigma} \ge \frac{a}{\tau}$ implies that $\Phi\left(\frac{a}{\sigma}\right) \ge \Phi\left(\frac{a}{\tau}\right)$ and $\frac{-a}{\sigma} \le \frac{-a}{\tau}$ implies that $\Phi\left(\frac{-a}{\sigma}\right) \le \Phi\left(\frac{-a}{\tau}\right)$, it follows that

$$\Phi\left(\frac{a}{\sigma}\right) - \Phi\left(-\frac{a}{\sigma}\right) \ge \Phi\left(\frac{a}{\tau}\right) - \Phi\left(-\frac{a}{\tau}\right).$$

So at least for estimators that are unbiased and asymptotically normal the variance completely determines the optimal estimator. This leads to consideration, for finite n, of the unbiased estimator of minimum variance even though it may not fulfill the close in probability frequency specification.

7.3   ESTIMATION ERROR BOUNDS

We now present a key theorem for the error of estimation.

Theorem 7.1   Let $X = (X_1, \ldots, X_n)$ and $f_X(x|\theta)$ be the generalized density with respect to σ-finite measure μ. Let $T(X)$ be any unbiased estimator of $\tau(\theta)$, then for θ and θ'

$$\text{var}(T|\theta) \ge \sup_{\theta'} \frac{[\tau(\theta') - \tau(\theta)]^2}{\text{var}\left(\frac{f(x|\theta')}{f(x|\theta)}\Big|\theta\right)} \quad \text{where } E_{\theta'}(T) = \tau(\theta').$$

Proof: We have

$$E_\theta\left[T\left(\frac{f(x|\theta') - f(x|\theta)}{f(x|\theta)}\right)\right] = \int\left(T\frac{f_{\theta'}}{f_\theta}f_\theta - Tf_\theta\right)d\mu = \tau(\theta') - \tau(\theta).$$

Using $\text{Cov}(Z,W) = E(ZW) - E(Z)E(W)$,

$$\text{Cov}\left\{T,\left(\frac{f(x|\theta')}{f(x|\theta)} - 1\right)\right\} = E\left\{(T - \tau)\left(\frac{f(x|\theta')}{f(x|\theta)} - 1\right)\Big|\theta\right\} = \tau(\theta') - \tau(\theta),$$

since

$$E_\theta\left(\frac{f(x|\theta')}{f(x|\theta)} - 1\right) = \int f(x|\theta')d\mu - 1 = 1 - 1 = 0.$$

Now recall the Cauchy-Schwarz inequality,

$$E^2(h(y)g(y)) \leq (Eh^2)(Eg^2)$$

with equality iff $h = cg$. Let $h = T - \tau(\theta)$, $g = f_{\theta'}/f_\theta - 1$. Then

$$0 \leq E^2(T - \tau)\left(\frac{f_{\theta'}}{f_\theta} - 1\right) = (\tau(\theta') - \tau(\theta))^2 \leq \mathrm{var}(T|\theta) \times \mathrm{var}\left(\frac{f(x|\theta')}{f(x|\theta)}|\theta\right),$$

so that

$$\mathrm{var}(T|\theta) \geq \sup_{\theta'} \frac{[\tau(\theta') - \tau(\theta)]^2}{\mathrm{var}\left(\frac{f(x|\theta')}{f(x|\theta)}|\theta\right)}.$$

□

Critical regularity conditions have not been assumed yet, but now let $d\tau(\theta)/d\theta$ exist and assume

$$\mathrm{var}\left(\frac{f_{\theta'}}{f_\theta}|\theta\right)\bigg/(\theta' - \theta)^2 \xrightarrow[\theta' \to \theta]{} J(\theta),$$

exists. Then

$$\mathrm{var}(T|\theta) \geq \lim_{\theta' \to \theta}\left[\left(\frac{\tau(\theta') - \tau(\theta)}{\theta' - \theta}\right)^2 \div \left\{\mathrm{var}\left(\frac{f_{\theta'}}{f_\theta}|\theta\right)\bigg/(\theta' - \theta)^2\right\}\right] = [\tau'(\theta)]^2/J(\theta).$$

Further, let $\frac{df(x|\theta)}{d\theta}$ exist, and for $|\theta' - \theta| < \epsilon$ (a fixed θ independent of X) let

$$(i) \qquad \left|\frac{f_{\theta'} - f_\theta}{(\theta' - \theta)f_\theta}\right| < G(x, \theta)$$

hold such that $E_\theta(G^2)$ exists which also implies $E_\theta(G)$ exists. Now

$$J(\theta) = \lim_{\theta' \to \theta} \frac{\mathrm{var}\left(\frac{f_{\theta'}}{f_\theta}|\theta\right)}{(\theta' - \theta)^2} = \lim_{\theta' \to \theta} \int \left(\frac{f_{\theta'} - f_\theta}{f_\theta}\right)^2 \frac{f_\theta}{(\theta' - \theta)^2}d\mu.$$

Using the Lebesgue dominated convergence theorem,

$$J(\theta) = \int \lim_{\theta' \to \theta} \left[\frac{f_{\theta'} - f_\theta}{\theta' - \theta} \right]^2 \frac{f_\theta}{f_\theta^2} d\mu = \int \left(\frac{\partial f_\theta}{\partial \theta} \right)^2 \frac{1}{f_\theta^2} f_\theta d\mu$$

$$= \int \left(\frac{\partial \log f_\theta}{\partial \theta} \right)^2 f_\theta d\mu = E \left(\frac{\partial \log f_\theta}{\partial \theta} \right)^2 = I(\theta),$$

the Fisher Information quantity. Hence the right-hand side of

$$\text{var}(T) \geq (\tau'(\theta))^2 / I(\theta)$$

is referred to as the Cramér-Rao lower bound, which in this "regular" case is the minimum variance bound for all T unbiased for $\tau(\theta)$. Note that condition (i) implies that

$$0 = \frac{\partial}{\partial \theta} \int f d\mu = \int \frac{\partial f}{\partial \theta} d\mu = \int \frac{\partial \log f}{\partial \theta} f d\mu = E \left(\frac{\partial \log f}{\partial \theta} \right).$$

Suppose a second derivative can be passed through the integral sign. Then

$$0 = \frac{\partial}{\partial \theta} \int \frac{\partial \log f}{\partial \theta} f d\mu = \int \left[f \frac{\partial^2 \log f}{\partial \theta^2} + \left(\frac{\partial \log f}{\partial \theta} \right) \frac{\partial f}{\partial \theta} \right] d\mu$$

$$= \int \frac{\partial^2 \log f}{\partial \theta^2} f d\mu + \int \left(\frac{\partial \log f}{\partial \theta} \right)^2 f d\mu = E \frac{\partial^2 \log f}{\partial \theta^2} + I(\theta),$$

or

$$I(\theta) = -E \left(\frac{\partial^2 \log f}{\partial \theta^2} \right).$$

Now in the regular case,

$$\tau(\theta) = \int T f_\theta d\mu = E(T),$$

$$\tau'(\theta) = \int T \frac{\partial f_\theta}{\partial \theta} d\mu = \int T \frac{\partial \log f_\theta}{\partial \theta} f_\theta d\mu = E \left(T \frac{\partial \log f_\theta}{\partial \theta} \right).$$

By the Cauchy-Schwarz inequality,

$$[\tau'(\theta)]^2 \leq \text{var}(T) E \left(\frac{\partial \log f_\theta}{\partial \theta} \right)^2$$

with equality iff

$$\frac{\partial \log f_\theta}{\partial \theta} = A(\theta)(T - \tau(\theta)),$$

and the lower bound is achieved. Note also in this case

$$E\left(\frac{\partial \log f_\theta}{\partial \theta}\right)^2 = A^2(\theta)\mathrm{var}(T),$$

and from the previous work,

$$E\left(\frac{\partial \log f_\theta}{\partial \theta}\right)^2 = \frac{[\tau'(\theta)]^2}{\mathrm{var}(T)}.$$

Then

$$A^2\mathrm{var}(T) = (\tau'(\theta))^2/\mathrm{var}(T),$$

or

$$\mathrm{var}(T) = \frac{\tau'(\theta)}{A(\theta)},$$

and if $\tau(\theta) = \theta$ then

$$\mathrm{var}(T) = \frac{1}{A(\theta)}.$$

If

$$\frac{\partial \log f}{\partial \theta} \neq A(\theta)(T - \tau),$$

the minimum bound will not be attained but there still may be an attainable minimum variance though larger than the regular minimum bound. If the "regularity"
conditions do not hold the smallest attainable variance may be less than the "regular
minimum bound,"

$$[\tau'(\theta)]^2/I(\theta).$$

Finally, note that $\tau'(\theta)$ and $A(\theta)$ must have the same sign, so there is no need for
absolute value in the above variance formula.

Example 7.3

For a non-regular case consider $X_1, \ldots, X_n$ i.i.d. with

$$f_\theta = f(x|\theta) = \begin{cases} e^{\theta - x} & \text{for } x \geq \theta \\ 0 & \text{for } x < \theta. \end{cases}$$

Since $\partial f_\theta / \partial \theta$ does not exist, $\int f_\theta dx = 1$ cannot be differentiated. The estimate, $\theta^* = \min_i (x_i) - \frac{1}{n}$, is unbiased and has variance less than the Cramér lower bound for $n > 1$.

If $\tau(\theta) = \theta$ then under regularity conditions

$$\text{var}(T) \geq \frac{1}{I(\theta)}.$$

This is a special case of the Cramér-Rao lower bound. When the minimum bound is not attained one can still find lower bounds, Bhattacharyya (1940).

Let

$$D_s = T - \tau(\theta) - \sum_{j=1}^{s} a_j \frac{1}{f_\theta} \frac{\partial^j f_\theta}{\partial \theta^j} = T - \tau(\theta) - a'\mathcal{L},$$

$$a' = (a_1, \ldots, a_s), \quad \mathcal{L}' = \frac{1}{f_\theta} \left(\frac{\partial f_\theta}{\partial \theta}, \frac{\partial^2 f_\theta}{\partial \theta^2}, \ldots, \frac{\partial^s f_\theta}{\partial \theta^s} \right) = (\mathcal{L}_1, \ldots, \mathcal{L}_s).$$

Assuming regularity conditions

$$0 = \frac{\partial^j \int f_\theta d\mu}{\partial \theta^j} = \int \frac{\partial^j f_\theta}{\partial \theta^j} d\mu = \int \frac{1}{f_\theta} \frac{\partial^j f_\theta}{\partial \theta^j} f_\theta d\mu$$

$$= E\left[\frac{1}{f_\theta} \frac{\partial^j f_\theta}{\partial \theta^j} \right] = E(\mathcal{L}_j),$$

$$E(D_s) = E(T - \tau(\theta)) - a' E(\mathcal{L}) = 0,$$

$$\text{var}(D_s) = E(D_s^2) = E(T - \tau(\theta) - a'\mathcal{L})^2$$

$$= E(T - \tau(\theta))^2 - 2a' E\mathcal{L}(T - \tau(\theta)) + a'(E\mathcal{L}\mathcal{L}')a$$

$$= \text{var}(T) - 2a'[E\mathcal{L}T - E\mathcal{L}\tau(\theta)] + a'Ba, \quad B = E(\mathcal{L}\mathcal{L}')$$

$$= \text{var}(T) - 2a' E\mathcal{L}T + a'Ba.$$

Now

$$E(T\mathcal{L}_j) = E\left(T \frac{1}{f_\theta} \frac{\partial^j f_\theta}{\partial \theta^j} \right) = \int T \frac{\partial^j f_\theta}{\partial \theta^j} d\mu = \frac{\partial^j}{\partial \theta^j} \int T f_\theta d\mu = \frac{\partial^j}{\partial \theta^j} \tau(\theta) = \gamma_j.$$

Hence

$$\mathrm{var}(D_s) = \mathrm{var}(T) - 2a'\gamma + a'Ba,$$

where

$$\gamma' = (\gamma_1, \ldots, \gamma_s),$$
$$\mathrm{var}(D_s) = \mathrm{var}(T) + (a - B^{-1}\gamma)'B(a - B^{-1}\gamma) - \gamma'B^{-1}\gamma \geq 0,$$
$$\min_a \mathrm{var}(D_s) = \mathrm{var}(T) - \gamma'B^{-1}\gamma \geq 0$$

that is, when $a = B^{-1}\gamma$

$$\mathrm{var}(T) \geq \gamma'B^{-1}\gamma.$$

Now $\mathrm{var}(D_s) = 0$ iff $D_s = 0 = T - \tau(\theta) - a'\mathcal{L}$ or $T - \tau(\theta) = a'\mathcal{L} = \gamma'B^{-1}\mathcal{L}$. Only in this case is the bound attained that is,

$$\mathrm{var}(T) = \gamma'B^{-1}\gamma.$$

One can show that

$$\mathrm{var}(T) \geq \gamma'B^{-1}\gamma \geq \frac{(\tau'(\theta))^2}{I(\theta)}.$$

More generally, for

$$\gamma' = \begin{pmatrix} \gamma^{(1)'}, & \gamma^{(2)'} \\ r & s-r \end{pmatrix} \quad \text{and} \quad B^{-1} = \begin{pmatrix} & r & s-r \\ B^{11} & B^{12} \\ B^{21} & B^{22} \end{pmatrix} \begin{matrix} r \\ s-r \end{matrix}$$

$$\mathrm{var}(T) \geq \gamma'B^{-1}\gamma \geq \gamma^{(1)'}B_{11}^{-1}\gamma^{(1)} + (\gamma^{(2)} - B_{21}B_{11}^{-1}\gamma^{(1)})'B^{22}(\gamma^2 - B_{21}B_{11}^{-1}\gamma^{(1)}).$$
If $\mathrm{var}(T) = \gamma^{(1)'}B_{11}^{-1}\gamma^{(1)}$ then there is no point going beyond $r = s$. In particular, if $r = 1$,

$$\mathrm{var}(T) \geq \gamma_1^2 b_{11}^{-1} + \left(\gamma^{(2)} - \frac{B_{21}}{b_{11}}\gamma_1\right)' B^{22} \left(\gamma^{(2)} - \frac{B_{21}}{b_{11}}\gamma_1\right),$$

and $\gamma_1^2 b_{11}^{-1} = (\tau'(\theta))^2/I(\theta)$. Note also the covariance matrix of $(T, \mathcal{L}')$ is

$$\mathrm{Cov}(T, \mathcal{L}') = \begin{pmatrix} \mathrm{var}(T) & \gamma' \\ \gamma & B \end{pmatrix}.$$

Therefore, the multiple correlation between T and $\mathcal{L}$ is $R^2_{T,\mathcal{L}} \equiv \frac{\gamma B^{-1} \gamma}{\text{var}(T)}$.

We now ask whether an unbiased estimator of minimum variance (if it exists) is unique irrespective of whether a minimum variance bound is attained. Let T and T^* both be unbiased and of minimum variance.

$$E(T) = E(T^*) = \tau(\theta) \quad \text{var}(T) = \text{var}(T^*) = \sigma^2.$$

Let

$$T' = \frac{1}{2}(T + T^*), \quad E(T') = \tau(\theta),$$

$$\text{var}(T') = \frac{1}{4}(\text{var}(T) + 2\,\text{Cov}(T, T^*) + \text{var}(T^*)).$$

Now

$$\text{Cov}(T, T^*) = E(T - \tau)(T^* - \tau) \le \sqrt{\text{var}(T)\,\text{var}(T^*)} = \sigma^2,$$

by the Cauchy-Schwarz inequality such that

$$\text{var}(T') \le \frac{1}{4}(\sigma^2 + 2\sigma^2 + \sigma^2) = \sigma^2.$$

Now the inequality sign contradicts the assumption that σ^2 is the minimum variance. However equality of $\text{Cov}(T, T^*) = \sqrt{\text{var}(T)\,\text{var}(T^*)} = \sigma^2$ implies that $(T - \tau) = u(\theta)(T^* - \tau)$ and $\text{Cov}(T, T^*) = E(T - \tau)(T^* - \tau) = u(\theta)E(T^* - \tau)^2 = u(\theta)\sigma^2$ but $\text{Cov}(T, T^*) = \sigma^2$ such that $u(\theta) \equiv 1$ and $T = T^*$ thus the minimum variance unbiased estimator is unique if it exists.

7.4 EFFICIENCY AND FISHER INFORMATION

Suppose for a given sample size n, T'_n has minimum variance among all unbiased estimators of θ. Then the efficiency of any unbiased statistic T_n can be defined as

$$0 \le e(T_n) = \frac{\text{var}(T'_n)}{\text{var}(T_n)} \le 1,$$

and $\lim_{n\to\infty} e(T_n)$ is termed the asymptotic efficiency.

In the regular case where the lower bound is $\text{var}(T_n) \geq \frac{1}{I_n(\theta)}$ and $X_1, X_2, \ldots,$ are i.i.d. from $f_X(x|\theta)$, where

$$I_n(\theta) = E\left(\frac{\partial \log \prod_{i=1}^n f(x_i|\theta)}{\partial \theta}\right)^2 = E\left(\frac{\partial \sum_i \log f(x_i|\theta)}{\partial \theta}\right)^2$$

$$= E\left(\sum_i \frac{\partial \log f(x_i|\theta)}{\partial \theta}\right)^2 = \sum_i E\left(\frac{\partial \log f(x_i|\theta)}{\partial \theta}\right)^2 = nI(\theta),$$

where $I(\theta) = E\left(\frac{\partial \log f(x|\theta)}{\partial \theta}\right)^2$ that is, of a single observation because $X_1, \ldots, X_n$ are i.i.d.

Hence another definition for the i.i.d. regular case is

$$e(T_n) = \frac{1}{nI(\theta)\,\text{var}(T_n)}.$$

If

$$\frac{\partial \log \prod_{i=1}^n f(x_i|\theta)}{\partial \theta} = \sum_{i=1}^n \frac{\partial \log f(x_i|\theta)}{\partial \theta} = A(\theta)(T_n - \tau),$$

then T_n is fully efficient. Another way of looking at efficiency is

$$0 \leq e_f = \frac{I_{T_n}(\theta)}{I_n(\theta)} \leq 1 \quad \text{where } I_{T_n}(\theta) = E\left(\frac{\partial \log f(t_n|\theta)}{\partial \theta}\right)^2.$$

We show that $I_{T_n}(\theta) \leq I_n(\theta)$.

For any measurable set $C \subset S$, the sample space of $(X_1, \ldots, X_n) = X$, let

$$\frac{d}{d\theta}P[X \in C] = \frac{d}{d\theta}\int_C f(x|\theta)d\mu = \int_C \frac{\partial \log f(x|\theta)}{\partial \theta}f(x|\theta)d\mu,$$

and let T_n be a statistic with probability function $g_{T_n}(t_n|\theta)$, satisfying

$$\frac{d}{d\theta}P[T_n \in C'] = \frac{d}{d\theta}\int_{C'} g(t_n|\theta)dv = \int_{C'} \frac{\partial \log g(t_n|\theta)}{\partial \theta}g(t_n|\theta)dv$$

for all $C' \subset S'$, the space of $T_n(X)$. Let $E\left(\frac{\partial \log g(t_n|\theta)}{\partial \theta}\right)^2 = I_{T_n}(\theta)$, then $I_n(\theta) \geq I_{T_n}(\theta)$.

Proof: We have

$$0 \leq E\left[\frac{\partial \log f(x|\theta)}{\partial \theta} - \frac{\partial \log g(t_n|\theta)}{\partial \theta}\right]^2 = I_n(\theta) + I_{T_n}(\theta) - 2E\left[\frac{\partial \log f}{\partial \theta} \cdot \frac{\partial \log g}{\partial \theta}\right],$$

and for $C' \subset S'$, there is a corresponding $C \subset S$ so that

$$P(T_n \in C'|\theta) = P(X \in C|\theta)$$

and

$$\int_{C'} \frac{\partial \log g}{\partial \theta} g dv = \int_C \frac{\partial \log f}{\partial \theta} f d\gamma.$$

Since this holds for every C it satisfies the definition of conditional expectation (see the Radon-Nykodym Theorem), that is,

$$\frac{\partial \log g}{\partial \theta} = E\left[\frac{\partial \log f}{\partial \theta} | T_n = t\right]$$

and

$$E_{T_n} E_{X|T_n}\left[\frac{\partial \log f}{\partial \theta} \frac{\partial \log g}{\partial \theta}\right] = E_{T_n}\left[\frac{\partial \log g}{\partial \theta}\right]^2 = I_{T_n}(\theta) \quad \text{and}$$

$$0 \le I_n(\theta) + I_{T_n}(\theta) - 2I_{T_n}(\theta) = I_n(\theta) - I_{T_n}(\theta) \quad \text{and} \quad I_n(\theta) \ge I_{T_n}(\theta),$$

as required.   □

7.5 INTERPRETATIONS OF FISHER INFORMATION

We consider the likelihood $L(\theta|D)$ or simply $L(\theta)$. Recall that $L(\theta)$ was useful for discriminating between alternative values of θ say θ_1 and θ_2 and if $L(\theta_1) = L(\theta_2)$ we cannot discriminate between the support of θ_1 and θ_2 for the given data. More generally if $L(\theta) = Constant$ for all θ there is obviously no information in the sample with regard to θ. So information in the sample with regard to θ should be associated with changes in $L(\theta)$ for varying θ. Now $\frac{\partial L}{\partial \theta}$ is the rate of change of L with respect to θ. Further, the greater L the less important is the same rate $\frac{\partial L}{\partial \theta}$. So it seems reasonable to consider the relative rate of change that is,

$$\frac{1}{L}\frac{\partial L}{\partial \theta} = \frac{\partial \log L}{\partial \theta} = S(\theta),$$

often called the Score function. Now when the regularity conditions (interchange of partial derivatives and integrals) hold,

$$E(S) = E\left(\frac{\partial \log L}{\partial \theta}\right) = 0 \text{ mean rate of relative change.}$$

It is reasonable to assume that the more variation in S, the more information in the sample relative to θ. Therefore since a measure of variation is the variance we may define

$$\text{var}(S) = E\left(\frac{\partial \log L}{\partial \theta}\right)^2,$$

as the average amount of information in the data relative to θ.

Another interpretation is curvature or how steep is the descent about the maximum. The curvature of $\log L(\theta)$ is defined as

$$C = \lim_{\Delta\theta \to 0} \frac{\Delta\phi}{\Delta s} = \frac{\dfrac{\partial^2 \log L(\theta)}{\partial \theta^2}}{\left[1 + \dfrac{\partial \log L(\theta)}{\partial \theta}\right]^{3/2}}.$$

For $\theta = \hat{\theta}$ since $L(\hat{\theta})$ maximizes $L(\theta)$ then $\log L(\hat{\theta})$ maximizes $\log L(\theta)$ so $\frac{\partial \log L(\theta)}{\partial \theta}|_{\theta=\hat{\theta}} = 0$ and

$$C = \frac{\partial^2 \log L(\theta)}{\partial \theta^2}|_{\theta=\hat{\theta}}$$

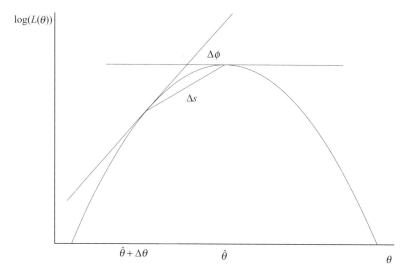

Figure 7.1

is the curvature at the maximum. Since this is negative it measures how steep the descent is around the Maximum Likelihood Estimate. Geometrically, the curvature at $\hat{\theta}$ is defined to be the rate of change of the angle of the tangent at $\hat{\theta} + \Delta\theta$ as $\Delta\theta$ tends to zero (see Figure 7.1). As already noted under certain regularity conditions

$$E\left[\frac{\partial \log L(\theta)}{\partial \theta}\right]^2 = -E\left[\frac{\partial^2 \log L(\theta)}{\partial \theta^2}\right].$$

Further interpretations:
Suppose we let

$$D(\theta', \theta) = E_{\theta'}\varphi\left(\frac{f_\theta}{f_{\theta'}}\right) \geq 0$$

be a discrepancy measure of f_θ from $f_{\theta'}$, where $\varphi(\cdot)$ is a convex function and $\varphi(1) = 0$ that is, $D(\theta',\theta) = 0$ for $\theta = \theta'$. Assume we can differentiate under the integral sign with respect to θ', then

$$\frac{\partial D}{\partial \theta'} = D'(\theta', \theta) = \frac{\partial}{\partial \theta'}\int \varphi f_{\theta'}d\mu = \int\left(f_{\theta'}'\varphi - f_{\theta'}\frac{\varphi'f_\theta'f_\theta}{f_{\theta'}^2}\right)d\mu$$

$$= \int\left[f_{\theta'}'\varphi - \frac{\varphi'f_\theta f_{\theta'}'}{f_{\theta'}}\right]d\mu,$$

where $\partial\varphi/\partial z = \varphi'$, $\partial\varphi/\partial\theta' = \varphi'f_\theta f_{\theta'}'/(-f_{\theta'}^2)$.
    Further,

$$\frac{\partial D'(\theta', \theta)}{\partial \theta'} = D''(\theta', \theta) = \int f_{\theta'}''\varphi - \frac{f_{\theta'}' \cdot \varphi'f_{\theta'}'f_\theta}{f_{\theta'}^2}$$

$$-\left[-\frac{\varphi''f_\theta^2(f_{\theta'}')^2}{f_{\theta'}^3} + \varphi'f_\theta\frac{\partial^2 \log f_{\theta'}}{\partial\theta^2}\right]d\mu.$$

Now $\frac{\partial^2 \log f_{\theta'}}{\partial\theta'^2} = \frac{f_{\theta'}''}{f_{\theta'}} - \frac{(f_{\theta'}')^2}{f_{\theta'}^2}$, so

$$D''(\theta', \theta) = \int\left[f_{\theta'}''\varphi + \frac{\varphi''f_\theta^2(f_{\theta'}')^2}{f_{\theta'}^3} - \frac{\varphi'f_{\theta'}''f_\theta}{f_{\theta'}}\right]d\mu.$$

Let

$$\frac{\partial^2 D(\theta', \theta)}{\partial\theta^2}|_{\theta'=\theta} = D''(\theta, \theta).$$

Then

$$D''(\theta,\ \theta) = 0 + \varphi''(1) \int f_\theta \left(\frac{\partial \log f_\theta}{\partial \theta} \right)^2 d\mu - \varphi'(1) \int f_\theta'' d\theta.$$

Now $D(\theta',\theta) \geq 0$ and $D(\theta,\theta) = 0$ then $D(\theta',\theta)$ has a minimum when $\theta' = \theta$ and $\varphi(1) = 0$. Hence

$$D''(\theta,\ \theta) = \varphi''(1) \int \left(\frac{\partial \log f_\theta}{\partial \theta} \right)^2 f_\theta d\mu = \varphi''(1) I(\theta).$$

Since φ is convex, $\varphi''(1) \geq 0$ (Burbea and Rao, 1982). In particular, if we let $\varphi = -\log \frac{f_\theta}{f_{\theta'}}$ then since the Kullback-Leibler divergence is

$$K(\theta',\ \theta) = E_\theta\left(-\log \frac{f_\theta}{f_{\theta'}} \right) = \int f_\theta \log \frac{f_\theta}{f_{\theta'}} d\mu,$$

we note that

$$K''(\theta,\ \theta) = \varphi''(1) I(\theta) = I(\theta) \text{ since } \varphi' = -\frac{1}{\left(\frac{f_\theta}{f_{\theta'}} \right)},\ \varphi'' = \left(\frac{f_\theta}{f_{\theta'}} \right)^2,$$

so that $\varphi''(1) = 1$.

As another special case, consider the distance

$$D(\theta',\ \theta) = \int \left(f_{\theta'}^{\frac{1}{2}} - f_\theta^{\frac{1}{2}} \right)^2 d\mu = \int f_{\theta'} \left(1 - \left(\frac{f_\theta}{f_{\theta'}} \right)^{\frac{1}{2}} \right)^2 d\mu,$$

so

$$\varphi\left(\frac{f_\theta}{f_{\theta'}} \right) = \left[1 - \left(\frac{f_\theta}{f_{\theta'}} \right)^{\frac{1}{2}} \right]^2$$

$$\varphi''(1) = \frac{1}{2}$$

$$D''(\theta,\ \theta) = \frac{1}{2} I(\theta).$$

Since

$$\int \left(f_{\theta'}^{\frac{1}{2}} - f_{\theta}^{\frac{1}{2}} \right)^2 d\mu = 2 - 2 \int (f_{\theta} f_{\theta'})^{\frac{1}{2}} d\mu,$$

then

$$0 \le D(\theta', \theta) \le 2,$$

with lower bound achieved iff $f_{\theta'} = f_{\theta}$, and the upper bound when $f_{\theta} f_{\theta'} = 0$, Pitman (1979). So $I(\theta)$ or $\frac{1}{2}I(\theta)$ is basically the curvature at the minimum discrepancy and expresses how steep is the ascent for changes about θ.

7.6 THE INFORMATION MATRIX

Let X have probability function $f_X(x|\theta)$ where $\theta = (\theta_1, \ldots, \theta_p)$ and

$$\frac{\partial \log f}{\partial \theta} = \begin{pmatrix} \dfrac{\partial \log f}{\partial \theta_1} \\ \vdots \\ \dfrac{\partial f}{\partial \theta_p} \end{pmatrix},$$

then

$$\mathcal{I}(\underset{\sim}{\theta}) = E\left[\left(\frac{\partial \log f}{\partial \theta} \right) \left(\frac{\partial \log f}{\partial \theta} \right)' \right].$$

Suppose $X \sim N(\mu, \sigma^2)$ and $\theta_1 = \mu$, $\theta_2 = \sigma^2$, then

$$\mathcal{I}(\mu, \sigma^2) = \begin{pmatrix} \dfrac{1}{\sigma^2} & 0 \\ 0 & \dfrac{1}{2\sigma^4} \end{pmatrix}.$$

Now Jeffreys' criterion for the prior with minimal information is to use it as joint prior for μ and σ^2,

$$g(\mu, \sigma^2) \propto |\mathcal{I}(\mu, \sigma^2)|^{\frac{1}{2}} \propto \frac{1}{\sigma^3}.$$

However, this did not appeal to Jeffreys and he prefered to use $g(\mu, \sigma^2) \propto \frac{1}{\sigma^2}$ that is, μ is uniform and $\log \sigma^2$ is uniform—so there are problems with the square root of the determinant of the information matrix, although this keeps the prior invariant, as we now show.

Suppose

$$\tau_1 = \tau_1(\theta_1, \ldots, \theta_p) = \tau_1(\theta), \qquad \theta_1 = \theta_1(\tau_1, \ldots, \tau_p) = \theta_1(\tau)$$

$$\vdots \qquad\qquad\qquad\qquad\qquad \vdots$$

$$\tau_p = \tau_p(\theta_1, \ldots, \theta_p) = \tau_p(\theta), \qquad \theta_p = \theta_p(\tau_1, \ldots, \tau_p) = \theta_p(\tau),$$

where $\tau = (\tau_1, \ldots, \tau_p)$. Further,

$$\frac{\partial \log f}{\partial \tau_j} = \frac{\partial \log f}{\partial \theta_1} \cdot \frac{\partial \theta_1}{\partial \tau_j} + \frac{\partial \log f}{\partial \theta_2} \cdot \frac{\partial \theta_2}{\partial \tau_j} + \cdots + \frac{\partial \log f}{\partial \theta_p} \cdot \frac{\partial \theta_p}{\partial \tau_j} \quad j = 1, \ldots, p.$$

We can express this in matrix form as

$$\frac{\partial \theta}{\partial \tau} = \begin{pmatrix} \dfrac{\partial \theta_1}{\partial \tau_1} & \cdots & \cdots & \dfrac{\partial \theta_p}{\partial \tau_1} \\ & \cdot & \cdot & \\ \cdot & & \cdot & \\ \cdot & & & \cdot \\ \cdot & & \cdot & \\ \dfrac{\partial \theta_1}{\partial \tau_p} & \cdots & \cdots & \dfrac{\partial \theta_p}{\partial \tau_p} \end{pmatrix}$$

and

$$\frac{\partial \log f}{\partial \tau} = \begin{pmatrix} \dfrac{\partial \log f}{\partial \tau_1} \\ \vdots \\ \dfrac{\partial \log f}{\partial \tau_p} \end{pmatrix},$$

so that

$$\frac{\partial \log f}{\partial \tau} = \frac{\partial \theta}{\partial \tau} \frac{\partial \log f}{\partial \theta}.$$

Hence

$$\mathcal{I}(\tau) = E\left(\frac{\partial \log f}{\partial \tau}\right)\left(\frac{\partial \log f}{\partial \tau}\right)' = E\left(\frac{\partial \theta}{\partial \tau} \frac{\partial \log f}{\partial \theta}\right)\left(\frac{\partial \theta}{\partial \tau} \frac{\partial \log f}{\partial \theta}\right)'$$

$$= E\left[\left(\frac{\partial \theta}{\partial \tau}\right)\left(\frac{\partial \log f}{\partial \theta}\right)\left(\frac{\partial \log f}{\partial \theta}\right)'\left(\frac{\partial \theta}{\sim \tau}\right)'\right]$$

$$= \left(\frac{\partial \theta}{\partial \tau}\right)\mathcal{I}(\theta)\left(\frac{\partial \theta}{\partial \tau}\right)'.$$

We now note that

$$\frac{\partial \tau}{\partial \theta} = \begin{pmatrix} \dfrac{\partial \tau_1}{\partial \theta_1} & \cdot & \cdot & \cdot & \cdot & \dfrac{\partial \tau_p}{\partial \theta_1} \\ \cdot & & & & & \cdot \\ \cdot & & & & & \cdot \\ \cdot & & & & & \cdot \\ \dfrac{\partial \tau_1}{\partial \theta_p} & \cdot & \cdot & \cdot & \cdot & \dfrac{\partial \tau_p}{\partial \theta_p} \end{pmatrix}, \qquad \frac{\partial \theta}{\partial \tau} = \begin{pmatrix} \dfrac{\partial \theta_1}{\partial \tau_1} & \cdot & \cdot & \cdot & \cdot & \dfrac{\partial \theta_p}{\partial \tau_1} \\ \cdot & & & & & \cdot \\ \cdot & & & & & \cdot \\ \cdot & & & & & \cdot \\ \dfrac{\partial \theta_1}{\partial \tau_p} & \cdot & \cdot & \cdot & \cdot & \dfrac{\partial \theta_p}{\partial \tau_p} \end{pmatrix}$$

and

$$\frac{\partial \tau}{\partial \theta} \cdot \frac{\partial \theta}{\partial \tau} = A = \{a_{ij}\},$$

where

$$a_{ij} = \sum_{k=1}^{n} \frac{\partial \tau_k}{\partial \theta_i} \frac{\partial \theta_j}{\partial \tau_k} = \frac{\partial \theta_j}{\partial \theta_i} = \begin{cases} 0 & \text{if } i \neq j \\ 1 & \text{if } i = j \end{cases},$$

so that $A = I$ and

$$\frac{\partial \tau}{\partial \theta} = \left(\frac{\partial \theta}{\partial \tau} \right)^{-1}.$$

Therefore,

$$\left(\frac{\partial \theta}{\partial \tau} \right)^{-1} \mathcal{I}(\tau) \left(\frac{\partial \theta}{\partial \tau} \right)^{\prime -1} = \mathcal{I}(\theta)$$

or

$$\left(\frac{\partial \tau}{\partial \theta} \right) \mathcal{I}(\tau) \left(\frac{\partial \tau}{\partial \theta} \right)^{\prime} = \mathcal{I}(\theta).$$

Taking the determinant of both sides yields

$$\left| \frac{\partial \tau}{\partial \theta} \right|^2 |\mathcal{I}(\tau)| = |\mathcal{I}(\theta)|,$$

$$\left| \frac{\partial \tau}{\partial \theta} \right| |\mathcal{I}(\tau)|^{\frac{1}{2}} = |\mathcal{I}(\theta)|^{\frac{1}{2}}.$$

Hence, whether you start with τ or θ, as long as you use the square root of the information matrix, your posterior probability for τ will be the same if you start with θ and transform to τ or whether you started with τ.

Suppose T is an unbiased estimator of θ that is, $E(T) = \theta$ where expectation is taken componentwise. Let

$$Z = \begin{pmatrix} T \\ \dfrac{\partial \log f}{\partial \theta} \end{pmatrix}, \quad T = (T_1, \ldots, T_p), \quad f(x_1, \ldots, x_n | \theta) = f,$$

then

$$E\left(\frac{\partial \log f}{\partial \theta}\right) = 0,$$

and

$$\mathrm{Cov}(Z) = E\begin{pmatrix} T - \theta \\ \dfrac{\partial \log f}{\partial \theta} \end{pmatrix}\begin{pmatrix} T - \theta \\ \dfrac{\partial \log f}{\partial \theta} \end{pmatrix}' = \begin{pmatrix} \Sigma_T & I \\ I & \mathcal{I}(\theta) \end{pmatrix} = \Lambda,$$

is non-negative definite (nnd) since

$$E\left[(T_i - \theta_i)\frac{\partial \log f}{\partial \theta_j}\right] = E\left[T_i \frac{\partial \log f}{\partial \theta_j}\right] = \int t_i \frac{1}{f} \frac{\partial f}{\partial \theta_j} f d\mu$$

$$= \int t_i \frac{\partial f}{\partial \theta_j} d\mu = \frac{\partial}{\partial \theta_j} \int t_i f d\mu = \frac{\partial \theta_i}{\partial \theta_j} = \begin{cases} 0 & \text{if } i \neq j \\ 1 & \text{if } i = j \end{cases}.$$

Since Λ is nnd so is $A\Lambda A'$ for any real matrix A. Let

$$A = \begin{pmatrix} I & -\mathcal{I}^{-1}(\theta) \\ 0 & \mathcal{I}^{-1}(\theta) \end{pmatrix}$$

or if $\mathcal{I}^{-1}(\theta)$ does not exist use $\mathcal{I}^{-}(\theta)$, the pseudo-inverse. Then

$$A\Lambda A' = \begin{pmatrix} \Sigma_T - \mathcal{I}^{-1}(\theta) & 0 \\ 0 & \mathcal{I}^{-1}(\theta) \end{pmatrix},$$

since every principle minor of a nnd matrix is nnd then $\Sigma_T - \mathcal{I}^{-1}(\theta)$ is nnd for all T. Since all of the diagonal elements of a nnd matrix are non-negative we have that $var(T_j) \geq i^{jj}$ where i^{jj} is the jth diagonal element of $\mathcal{I}^{-1}(\theta)$, yielding a lower bound for the variance of the unbiased estimator of $\theta_j, j = 1, \ldots, p$.

7.7 SUFFICIENCY

Definition of a Sufficient Statistic: $T = T(D)$ is sufficient for the family of probability functions $f(D|\theta)$ if the distribution of $D|T$ is independent of θ that is, $f(D|T = t, \theta)$ is independent of θ for all t.

Theorem 7.2 (without proof): $T(D)$ is sufficient for θ iff $f(D|\theta) = g(t|\theta)h(D)$. This shows that the information in a sufficient statistic is the same as in the sample D since

$$\frac{\partial \log f(D|\theta)}{\partial \theta} = \frac{\partial \log g(t|\theta)}{\partial \theta},$$

so that $I_D(\theta) = I_T(\theta)$, assuming the existence of $I_D(\theta)$.

Note that this does not provide an estimate of θ since any one-to-one transformation of a sufficient statistic is also sufficient. If θ and T a sufficient statistic for θ are both scalars then T is unique up to a one-to-one transformation. Suppose there are two scalar sufficient statistics U and T for scalar θ. Then we have

$$f(t, u|\theta) = f(u|\theta) f(t|u) = f(t|\theta) f(u|t).$$

Then

$$\frac{f(u|\theta)}{f(t|\theta)} = \frac{f(u|t)}{f(t|u)} = g(u, t),$$

such that the left-hand side cannot be a function of θ. The only way this can occur is if u and t are related by $1 - 1$ transformation, independent of θ.

7.8 THE BLACKWELL-RAO RESULT

Suppose U is any estimator of $\tau(\theta)$ not necessarily unbiased, and T is a sufficient statistic such that $E(U) = E(T)$. Then for $h(T) = E_{U|T}(U|T)$, since T is sufficient, we can show that

$$E[U - \tau(\theta)]^2 \geq E(h(T) - \tau(\theta))^2.$$

Now

$$E(U) = E_T E_{U|T}(U) = E_T[h(T)],$$

$$E(U - \tau)^2 = E[U - h(T) + h(T) - \tau]^2$$

$$= E[U - h(T)]^2 + E[h(T) - \tau]^2 + 2E_T E_{U|T}(U - h(T))(h(T) - \tau)$$

$$= E[U - h(T)]^2 + E(h(T) - \tau)^2 + 0,$$

so that $E(U - \tau)^2 \geq E(h(T) - \tau)^2$ with equality iff $U = h(T)$.

So if U is unbiased for τ so is $h(T)$ and we have shown how to do better than a given U by an $h(T)$. In the case of U unbiased we have not shown that $h(T)$ is the minimum variance unbiased estimate (mvue), but we do know if one exists it is unique. Now we suppose that a complete sufficient statistic exists and we recall the definition of completeness. T is complete if

$$E[g(T)] = 0 \quad \text{iff} \quad g(T) \equiv 0.$$

Then every function of T is the unique mvue for its expectation because if T is complete then for any two U and V such that $E(U) = E(V) = \tau$ and $E(U|T) = h_1(T)$ and $E(V|T) = h_2(T)$,

$$E(h_1 - h_2) = 0 \quad \text{iff} \quad h_1 = h_2.$$

For the exponential family

$$f(x|\theta) = e^{k(x)P(\theta) + Q(\theta) + C(x)},$$

where the range of x is independent of θ. Let $X_1, \ldots, X_n$ be i.i.d. Then

$$L = \prod_{i=1}^{n} f(x_i|\theta) = \exp\left(\sum_{i=1}^{n} k(x_i)P(\theta) + nQ(\theta) + \sum_i C(x_i)\right).$$

It is clear that $T_n = \sum_{i=1}^{n} k(x_i)$ is sufficient.

Given the regularity conditions the Cramér-Rao bound is attained for T_n satisfying $E(T_n) = \tau(\theta)$ if

$$\frac{\partial \sum \log f(x_i|\theta)}{\partial \theta} = A_n(\theta)(T_n - \tau(\theta)),$$

or

$$\sum \log f(x_i|\theta) = T_n \int^\theta A_n(\theta')d\theta' - \int^\theta \tau(\theta')A_n(\theta')d\theta' + R(T_n)$$
$$\equiv T_n(x)P_n^*(\theta) - Q_n^*(\theta) + C_n^*(T_n(x)),$$

or

$$f(x|\theta) = \exp\left(T_n(x)P_n^*(\theta) - Q_n^*(\theta) + C_n^*(T_n(x))\right).$$

Conversely, if f is exponential,

$$\log L = \Sigma \log f(x_i|\theta) = P(\theta)\sum_{i=1}^n k(x_i) + nQ(\theta) + \sum_i c(x_i),$$

and

$$\frac{\partial \log L}{\partial \theta} = P'(\theta)\Sigma k(x_i) + nQ'(\theta) = P'(\theta)\left[\sum_{i=1}^n k(x_i) - \frac{-nQ'(\theta)}{P'(\theta)}\right],$$

which satisfies the condition. So the Cramér-Rao lower bound is satisfied for $T_n = \sum_i k(x_i)$, where $E(T_n) = \frac{-nQ'(\theta)}{P'(\theta)} = \tau(\theta)$, since $E\left(\frac{\partial \log L}{\partial \theta}\right) = 0$ under the regularity conditions.

7.9 BAYESIAN SUFFICIENCY

Suppose for data D

$$L(\theta|D) = f(D|\theta) = f_T(t(D)|\theta)h(D),$$

then T is sufficient for θ. Now if $g(\theta)$ is a prior probability function on θ so that the posterior probability function

$$p(\theta|D) \propto L(\theta|D)g(\theta) \propto f(t|\theta)h(D)g(\theta)$$

or

$$p(\theta|D) = \frac{f(t|\theta)g(\theta)}{\int f(t|\theta)g(\theta)d\theta} = p(\theta|t).$$

So the posterior depends on the data D only through T. In other words any inference made with T is equivalent to one made with D. As before the factorization of the likelihood is the necessary and sufficient condition for sufficiency. Bayesian sufficiency depends on whatever the posterior probability function depends upon. This, in turn, depends on the usual sufficient statistic T and whatever other elements are introduced by the prior $g(\theta)$. At any rate sufficiency then is not a critical concept for Bayesian inference.

7.10 MAXIMUM LIKELIHOOD ESTIMATION

If the likelihood for a set of hypotheses specified by $\theta \in \Theta$ is $L(\theta|D)$, then the best supported θ is $\hat{\theta}$ for

$$L(\hat{\theta}|D) \geq L(\theta|D) \quad \text{for} \quad \theta \in \Theta.$$

It is the best supported value in the light of the Likelihood approach and it would seem reasonable to say $\hat{\theta}$ is the best estimator of θ (i.e., best supported value but not necessarily that we have good reason to believe it will be close to the true value in the long run).

Suppose $L(\theta|D)$ is continuous and differentiable. Then if

$$\frac{\partial L(\theta|D)}{\partial \theta} = 0$$

yields the stationary values we look for the global maximum. Often it is more convenient to set

$$\frac{\partial \log L(\theta|D)}{\partial \theta} = 0$$

to find the stationary values. Suppose T is a sufficient statistic and so that

$$L(\theta|D) = g(t|\theta)h(D),$$

then

$$\frac{\partial \log L}{\partial \theta} = \frac{\partial \log g(t|\theta)}{\partial \theta} = 0,$$

so $\hat{\theta}$ will be a function of T but not necessarily a one-to-one function so it need not be sufficient itself. If the conditions for a minimum variance bound hold so that

$$\frac{\partial \log L}{\partial \theta} = A(\theta)(T - \tau(\theta)),$$

T is sufficient for θ. Then solutions and stationary points are found for

$$\frac{\partial \log L}{\partial \theta} = 0 = T - \tau(\hat{\theta})$$

by solving $T = \tau(\hat{\theta})$ and

$$\frac{\partial^2 \log L}{\partial \theta^2} = A'(\theta)(T - \tau(\theta)) - A(\theta)\tau'(\theta).$$

We recall that in this situation

$$\tau'(\theta) = \mathrm{var}(T)A(\theta).$$

Note if $\tau(\theta) = \theta$ then $\mathrm{var}(T) = A^{-1}(\theta)$ and

$$\frac{\partial^2 \log L}{\partial \theta^2} = A'(\theta)(T - \tau(\theta)) - A^2(\theta)\mathrm{var}(T),$$

such that

$$\left. \frac{\partial^2 \log L}{\partial \theta^2} \right|_{\theta=\hat{\theta}} = A'(\hat{\theta})(T - \tau(\hat{\theta})) - A^2(\hat{\theta})\mathrm{var}(T) = -A^2(\hat{\theta})\mathrm{var}(T) < 0$$

is sufficient for $\hat{\theta}$ to be a relative maximum. Therefore all solutions $\hat{\theta}$ of $T - \tau(\hat{\theta}) = 0$ are maxima of $L(\theta|D)$. But all well-behaved functions have a minimum between successive relative maxima. Since all the stationary values are maxima—there is no minimum between successive maxima and hence there is only one maximum. So if a minimum variance bound estimator exists it is given by the Maximum Likelihood Estimator (MLE) since $\hat{\tau} = T$. Further there is a one-to-one correspondence between the existence of a single sufficient statistic for θ and the existence of a minimum variance unbiased estimate that attains the minimum bound for some $\tau(\theta)$ in the regular case, since in both cases the necessary and sufficient condition for their existence is the exponential family of distributions.

Example 7.4 (Regular Case)

Suppose

$$L(\theta) = \binom{n}{r}\theta^r(1-\theta)^{u-r} \quad \log L = r\log\theta + (n-r)\log(1-\theta) + \log\binom{n}{r},$$

$$\frac{\partial \log L}{\partial \theta} = \frac{r}{\theta} - \frac{n-r}{1-\theta} = 0,$$

so $\hat\theta = \frac{r}{n}$ is sufficient and the MLE. Now

$$\frac{\partial \log L}{\partial \theta} = r\left(\frac{1}{\theta} + \frac{1}{1-\theta}\right) - \frac{n}{1-\theta} = \frac{r}{\theta(1-\theta)} - \frac{n}{1-\theta} = \frac{1}{\theta(1-\theta)}(r - n\theta)$$

$$= \frac{n}{\theta(1-\theta)}\left(\frac{r}{n} - \theta\right) = A(\theta)(T - \tau(\theta)), \quad \text{so} \quad \text{var}(T) = \frac{\theta(1-\theta)}{n}.$$

Example 7.5 (Non-Regular Case)

Let

$$X_1, \ldots, X_n \quad \text{be i.i.d.} \quad f_x(x|\theta) = e^{-(x-\theta)} \quad x \ge \theta$$

and arranging the observations in order of size

$$x_{(1)} \le x_{(2)} \le \cdots \le x_n,$$

$$L(\theta) = \prod_{t=1}^{n} f(x_i|\theta) = e^{-\Sigma(x_{(i)}-\theta)} \propto e^{n\theta} \quad \text{for} \quad \theta \le x_{(1)},$$

so that $L(\theta)$ depends only on $x_{(1)}$, which is sufficient. Note also that

$$e^{nx_{(1)}} \ge e^{n\theta} \quad \text{for all admissible } \theta,$$

and that $x_{(1)}$ is also the MLE.

Example 7.6 (MLE not sufficient)

Consider $(X_i, Y_i) \quad i = 1, \ldots, n$ from

$$f_{X,Y}(x,y) = e^{-(x\theta + \frac{y}{\theta})}, \quad x > 0, \quad y > 0,$$

$$L(\theta) = e^{-(\theta\Sigma x_i + \frac{1}{\theta}\Sigma y_i)}.$$

Hence $(\Sigma x_i, \Sigma y_i)$ is the bivariate sufficient statistic for scalar θ. Now

$$\frac{d \log L}{d\theta} = -\Sigma x_i + \frac{1}{\theta^2} \Sigma y_i,$$

and $\hat{\theta}^2 = \frac{\Sigma y_i}{\Sigma x_i}$ so $\hat{\theta} = \sqrt{\Sigma y_i / \Sigma x_i}$ is the MLE. But the MLE is not a sufficient statistic. We will come back to this problem later.

7.11   CONSISTENCY OF THE MLE

Let $X_1, \ldots, X_n$ i.i.d. $f(x|\theta)$ and $D = (x_1, \ldots, x_n)$.
    Let $\hat{\theta}$ be such that

$$\log L(\hat{\theta}|D) \geq \log L(\theta|D).$$

Let θ_0 be the true value of θ and consider $\frac{L(\theta|D)}{L(\theta_0|D)}$ and E_0 be the expectation for true $\theta = \theta_0$.
    Now for $\theta^* \neq \theta_0$ and $L(\theta^*|D) \neq L(\theta_0|D)$ and by Jensen's Inequality,

$$E_0 \left(\log \frac{L(\theta^*|D)}{L(\theta_0|D)} \right) \leq \log E_0 \left(\frac{L(\theta^*|D)}{L(\theta_0|D)} \right) = \log 1 = 0,$$

and

$$(i) \quad E_0 \left[\frac{\log L(\theta^*|D)}{n} \right] \leq E_0 \left[\frac{\log L(\theta_0|D)}{n} \right],$$

with strict inequality unless $\theta^* = \theta_0$. Now

$$\frac{\log L(\theta|D)}{n} = \frac{1}{n} \sum_{i=1}^{n} \log f(x_i|\theta)$$

is the sample average of n i.i.d. random variables, where

$$E_{\theta_0} \{ \log f(x_i|\theta) \} = \int \log f(x|\theta) dF(x|\theta_0).$$

Then, by the Strong Law of Large Numbers,

$$\frac{1}{n} \Sigma \log f(x_i|\theta) \xrightarrow{P} E_0 \frac{1}{n} \Sigma \log f(x_i|\theta).$$

By virtue of (i) for large n with probability 1

$$\frac{1}{n}\sum_i \log f(x_i|\theta^*) < \frac{1}{n}\sum_i \log f(x_i|\theta_0) \quad \text{for all} \quad \theta^* \neq \theta_0,$$

or

$$\left[\frac{1}{n}\sum_i \log f(x_i|\theta^*) - \frac{1}{n}\sum_i \log f(x_i|\theta_0)\right] = E_{\theta_0}\left[\log \frac{f(x|\theta^*)}{f(x)|\theta_0)}\right] \leq 0.$$

But $L(\hat{\theta}|D) \geq L(\theta_0|D)$ or $\log L(\hat{\theta}|D) \geq \log L(\theta_0|D)$, for all n. So we must have

$$P \lim_{n\to\infty}[\hat{\theta}_n = \theta_0] = 1.$$

7.12 ASYMPTOTIC NORMALITY AND "EFFICIENCY" OF THE MLE

Let $L(\theta) = \prod_1^n f(x_i|\theta)$ where $X_i, \ldots, X_n$ are i.i.d. $f(x|\theta)$, and assume

$$E\left(\frac{d \log L(\theta)}{d\theta}\right) = 0,$$

and that

$$E\left(\frac{d \log L(\theta)}{d\theta}\right)^2 = -E\left(\frac{d^2 \log L(\theta)}{d\theta^2}\right)$$

exists and is non-zero for all θ in an interval about and including $\theta = \theta_0$ the true value. Then the MLE

$$\hat{\theta} \to N\left(\theta_0, \frac{1}{E\left[\left(\frac{d \log L}{d\theta}\right)^2\right]}\right),$$

in law.

We expand $\frac{d \log L}{d\theta}$ in a Tayor Series about $\theta = \theta_0$,

$$\frac{d \log L(\theta)}{d\theta} = \left(\frac{d \log L(\theta)}{d\theta}\right)_{\theta=\theta_0} + (\theta - \theta_0)\left(\frac{d^2 \log L(\theta)}{d\theta}\right)_{\theta=\theta^*},$$

for $\theta^* \in (\theta, \theta_0)$. Now

$$0 = \frac{d \log L(\theta)}{d\theta}\bigg|_{\theta=\hat{\theta}} = \left(\frac{d \log L(\theta)}{d\theta}\right)_{\theta=\theta_0} + (\hat{\theta} - \theta_0)\left(\frac{d^2 \log L(\theta)}{d\theta^2}\right)_{\theta=\theta^*},$$

or

$$(\hat{\theta} - \theta_0) = -\left(\frac{d \log L(\theta)}{d\theta}\right)_{\theta=\theta_0} \bigg/ \left(\frac{d^2 \log L(\theta)}{d\theta^2}\right)_{\theta=\theta^*}.$$

Since $\log L(\theta) = \sum \log f(x_i|\theta)$, $E[d \log f(x_i|\theta)/d\theta] = 0$ and

$$-E\left[\frac{d^2 \Sigma \log f(x_i|\theta)}{d\theta^2}\right] = -E\left(\frac{d^2 \log L(\theta)}{d\theta^2}\right) = nI(\theta),$$

then

$$I(\theta) = -E\left(\frac{d^2 \log f(x|\theta)}{d\theta^2}\right),$$

and

$$(\hat{\theta} - \theta_0)\sqrt{I_n(\theta_0)} = \left(-\frac{d \log L}{d\theta}\right)_{\theta=\theta_0} \bigg/ \left[\sqrt{I_n(\theta_0)}\left(\frac{d^2 \log L}{d\theta^2}\right)_{\theta=\theta^*} \bigg/ nI(\theta_0)\right].$$

Since $\hat{\theta} \to \theta_0$ (almost surely),

$$\left(\frac{1}{n}\frac{d^2 \log L(\theta)}{d\theta^2}\right)_{\theta=\theta^*} \longrightarrow I(\theta_0).$$

Since

$$\frac{d \log L}{d\theta}\bigg|_{\theta=\theta_0} = \frac{d\Sigma \log f(x_i|\theta)}{d\theta}\bigg|_{\theta=\theta_0},$$

$$(\hat{\theta} - \theta_0)\sqrt{I_n(\theta_0)} \longrightarrow \frac{d \log L}{d\theta}\bigg|_{\theta=\theta_0} \bigg/ \sqrt{I_n(\theta_0)} \longrightarrow N(0, 1),$$

and thus

$$(\hat{\theta} - \theta_0)\sqrt{n} \longrightarrow N\left(0, \frac{1}{I(\theta_0)}\right)$$

(in law) or

$$\hat{\theta} \sim N\left(\theta_0, \ \frac{1}{nI(\theta_0)}\right),$$

(approximately) by the Central Limit Theorem. Hence $\hat{\theta}$ is asymptotically normal and efficient.

7.13 SUFFICIENCY PRINCIPLES

We will now introduce a further duo of statistical principles. Recall that $T(D)$ is sufficient for θ iff

$$f(D|\theta) = g(D)h(T, \theta).$$

1. *Restricted Sufficiency Principle (RSP)*: For $\mathcal{E} = (S, \mu, \Theta, f)$,

$$\mathrm{Inf}(\mathcal{E}, D) = \mathrm{Inf}(\mathcal{E}, D'),$$

if $T(D) = T(D')$ for $T(D)$ sufficient for θ.

2. *Unrestricted Sufficiency Principle (USP)*: If T is sufficient for θ then

$$\mathrm{Inf}(\mathcal{E}, D) = \mathrm{Inf}(\mathcal{E}_T, T),$$

where $\mathcal{E}_T = (S_T, \mu, \Theta, f_T)$, S_T represents the space of T, and f_T its probability function.

Now we show that RSP $\Leftrightarrow$ RLP. Assuming T is sufficient implies

$$f(D|\theta) = g(D)h(T, \theta).$$

If $T(D) = T(D')$,

$$f(D'|\theta) = g(D')h(T, \theta),$$

or

$$f(D|\theta)g(D') = f(D'|\theta)g(D),$$

so

$$f(D|\theta) = g(D, D')f(D'|\theta).$$

Apply RLP and obtain $\mathrm{Inf}(\mathcal{E}, D) = \mathrm{Inf}(\mathcal{E}, D')$, so RLP $\Rightarrow$ RSP. Conversely, if

$$f(D|\theta) = g(D, D')f(D'|\theta),$$

then

$$g(D|T, \theta)h(T|\theta) = g(D, D')g(D'|T, \theta)h(T|\theta),$$

when $T(D') = T(D)$. Then

$$\frac{g(D|T, \theta)}{g(D'|T, \theta)} = g(D, D'),$$

which is constant in θ. The only way this can happen is if $g(D|T,\theta)$ is free of θ, so T is sufficient for θ and applying RSP yields RLP.

We now can state the various implications regarding relationships among the various statistical principles in the following diagram:

$$
\begin{array}{ccccccc}
& & \mathrm{USP} & \rightarrow & \mathrm{RSP} & & \\
& \nearrow & \uparrow & \searrow & \updownarrow & \searrow & \\
\mathrm{RCP\ \&\ MEP} & \longleftrightarrow & \mathrm{ULP} & \rightarrow & \mathrm{RLP} & \rightarrow & \mathrm{MEP} \\
& \searrow & \downarrow & & & & \\
& & \mathrm{UCP} & \rightarrow & \mathrm{RCP.} & &
\end{array}
$$

Hence, whether you accept LP as such or not, you should if you basically accept SP and CP. However, classical statistics bases its inferential procedures on the Repeated Sampling Principle which can violate LP.

REFERENCES

Barnard, G. A. (1974). Can we all agree what we mean by estimation? *Utilitas Mathematica*, **6**, 3–22.

Bhattacharyya, A. (1946). On some analogues of the amount of information and their use in statistical estimation. *Sankhya*, **8**, 1–14, 201–218, 315–328.

Burbea, J. and Rao, C. R. (1982). Entropy differential metric, distance and divergent measures in probability spaces: A unified approach. *Journal of Multivariate Analysis*, **12**, 575–596.

Pitman, E. J. G. (1979). *Some Basic Theory for Statistical Inference*. Chapman and Hall, London.

CHAPTER EIGHT

Set and Interval Estimation

This final chapter lays the foundations for developing confidence intervals, fiducial intervals, and (Bayesian) probability intervals. Considerable attention is paid to methods based on conditioning on ancillary statistics and to methods based on Fisher's "fiducial" distribution. Conditions are developed for the formal equivalence of fiducial and Bayesian intervals. Methods for multiparameter problems are discussed and illustrated by the Fisher-Behrens problem for obtaining interval inferences for a difference in normal means with unequal variances, and the Fieller-Creasy problem for obtaining interval inferences for the ratio of two normal means based on bivariate normal data.

8.1 CONFIDENCE INTERVALS (SETS)

Suppose $X_i, \ldots, X_n$ are i.i.d. $f(x|\theta)$ and $a(X_1, \ldots, X_n)$ and $b(X_1, \ldots, X_n)$ are two functions of the observations such that, independent of θ,

$$P[a \leq \theta \leq b|\,\theta] \geq 1 - \alpha.$$

Thus the probability is at least $1 - \alpha$ that the random interval $[a, b]$ includes or covers θ. If a and b can so be chosen they are called the lower and upper confidence limits and $1 - \alpha$ is the confidence coefficient. More generally if we have a random set $E(X_1, \ldots, X_n)$ such that, independent of θ,

$$P[\theta \in E|\,\theta] \geq 1 - \alpha,$$

we have a $1 - \alpha$ confidence set for θ.

Suppose, in addition, $f(x|\theta)$ is absolutely continuous and there is a function $p(X_1, \ldots, X_n, \theta)$ defined at every point θ in an interval about the true value θ_0, and every point in the sample space. Also p is a continuous and monotonically

Modes of Parametric Statistical Inference, by Seymour Geisser
Copyright © 2006 John Wiley & Sons, Inc.

increasing (or decreasing) function of θ, and has a distribution that is independent of θ. Then let $[p_1, p_2]$ be an interval for which

$$P(p_1 \leq p \leq p_2 | \theta) = 1 - \alpha \quad \text{(independent of } \theta\text{)}.$$

Then if θ_0 is the true value of θ, the solutions a and b of the equation $p(x_1, \ldots, x_n, \theta) = p_i$, $i = 1, 2$ exist and provide a $1 - \alpha$ confidence interval for θ_0. So that if θ_0 is the true value and say p is monotonically increasing (or decreasing) function of θ

$$P[a < \theta_0 < b | \theta_0] = 1 - \alpha,$$

where a and b are solutions to $p(x_1, \ldots, x_n, \theta_0) = p_i$ $i = 1, 2$.

Here a and b are random variables and the interval $[a, b]$ has probability of $1 - \alpha$ of including the true value θ_0 and $[a, b]$ is the confidence interval.

Example 8.1

$X_1, \ldots, X_n$ i.i.d. $N(\mu, 1)$. Let $Z = (\overline{X} - \mu)\sqrt{n} \sim N(0, 1)$, independent of μ. Then

$$P[-z_{\alpha/2} \leq Z \leq z_{\alpha/2} | \mu] = P[-z_{\alpha/2} \leq (\overline{X} - \mu)\sqrt{n} \leq z_{\alpha/2}]$$

$$= P\left[\overline{X} - \frac{z_{\alpha/2}}{\sqrt{n}} \leq \mu \leq \overline{X} + \frac{z_{\alpha/2}}{\sqrt{n}}\right]$$

$$= \int_{-z_{\alpha/2}}^{z_{\alpha/2}} \frac{e^{-\frac{1}{2}x^2}}{\sqrt{2\pi}} \, dx = 1 - \alpha.$$

If $F(x|\theta)$ is absolutely continuous and monotonically increasing (or decreasing) function of θ, we can always find a confidence interval for θ. Let $p(X_1, \ldots, X_n | \theta) = -2\sum_i \log F(X_i | \theta)$. Recall that $Z = F(X|\theta)$ has a uniform distribution, $0 \leq Z \leq 1$, so

$$F(x) = P(X \leq x) = P(F(X) \leq F(x)).$$

Letting $Z = F(X)$, and $Y = -2 \log Z$,

$$P(Y > y) = P(-2 \log Z > y) = P(Z \leq e^{-\frac{1}{2}y}) = e^{-\frac{1}{2}y}$$

or

$$f_Y(y) = \frac{1}{2} e^{-\frac{1}{2}y},$$

thus $Y \sim \chi_2^2$, a chi-square random variable with 2 degrees of freedom. Hence $Y_i = -2 \log F(X_i | \theta)$, and $p = \Sigma Y_i$ where Y_i are i.i.d. χ_2^2 variates so that $p \sim \chi_{2n}^2$ independent of θ, and

$$P[p_1 < p(X_1, \ldots, X_n | \theta) < p_2] = 1 - \alpha,$$

or

$$P\left[\chi_{2n}^2(\alpha/2) \leq -2 \sum_1^n \log F(X_i | \theta) \leq \chi_{2n}^2\left(\frac{1-\alpha}{2}\right)\right] = 1 - \alpha.$$

Solving for θ then provides a confidence interval. This demonstrates, for absolutely continuous $F(x|\theta)$ which is strictly monotonic in θ, that we can always find a confidence interval. But if $F(x|\theta)$ is not monotonic in θ, we can find a confidence set instead of an interval.

If we have scalar statistic T and its distribution $F(t|\theta)$ depends only on θ where $t' \leq t \leq t''$ and $\theta' \leq \theta \leq \theta''$, such that $F(t|\theta)$ is monotonically decreasing in θ, we can find a confidence interval.

Let

$$\int_{t'}^{t_1(\theta)} f(t|\theta)\, dt = \alpha/2 = F(t_1(\theta)|\theta),$$

$$\int_{t_2(\theta)}^{t''} f(t|\theta)\, dt = \alpha/2 = 1 - F(t_2(\theta)|\theta),$$

so that

$$F(t_2(\theta)|\theta) - F(t_1(\theta)|\theta) = \int_{t_1(\theta)}^{t_2(\theta)} f(t|\theta)\, dt = 1 - \alpha.$$

Hence

$$P(\theta_1(t_0) \leq \theta_0 \leq \theta_2(t_0)|\theta_0) = 1 - \alpha.$$

8.2 CRITERIA FOR CONFIDENCE INTERVALS

Based on statistic T_1 or T_2 we denote intervals $I(T_1)$ and $I(T_2)$, where

$$P[\theta \in I(T_1)|\theta] = 1 - \alpha = P[\theta \in I(T_2)|\theta].$$

If for every θ, for given $1 - \alpha$ and $L(T)$ being the length of the interval,

$$E[L(T_1)|\theta] \leq E[L(T_2)|\theta],$$

and less for at least one θ then T_1 is preferable to T_2. Similarly if one desires the shortest $L(T)$ that is,

$$\min_{T} E\{L(T)|\theta\},$$

this may generally be rather difficult to execute.

Another criterion is a selective one that is, T_1 is more selective than T_2 iff for every θ and $1 - \alpha$

$$P(\theta' \in I(T_1)|\theta) \leq P(\theta' \in I(T_2)|\theta),$$

and less for one θ', where θ' is any value of θ other than the true value. The most selective one is that T_1 for which

$$P(\theta' \in I(T_1)|\theta) < P(\theta' \in I(T)|\theta)$$

for any other T. In other words, $I(T_1)$ covers false values of θ least frequently. However these don't exist too often since they are constructed by inverting the acceptance interval of a UMP test—which, when available, is usually one sided.

Another criterion is a selective unbiased criterion. The probability that $I(T)$ contains θ when it is true is at least as large as containing any false θ'

$$P(\theta \in I(T)|\theta) \geq P(\theta' \in I(T)|\theta).$$

8.3 CONDITIONING

Fisher stressed the importance of conditioning in the presence of a relevant subset. A simple example is given that demonstrates its value and importance.

Example 8.2

Suppose X_1 and X_2 are independent and

$$P(X = \theta - 1) = P(X = \theta + 1) = \frac{1}{2} \quad \theta \in R.$$

In other words, X_i is as likely to be 1 unit above or below θ. Here a confidence set is a point where

$$C(x_1, x_2) = \begin{cases} \bar{x} & \text{if } x_1 \neq x_2 \\ x - 1 & \text{if } x_1 = x_2 = x \end{cases}$$

Now if $x_1 \neq x_2$, then $\theta \equiv \bar{x}$. But

$$P(X_1 \neq X_2|\theta)P(\overline{X} = \theta|X_1 \neq X_2) = \frac{1}{2} \times 1 = .5$$

and

$$P(X_1 = X_2 = X|\theta)P(X - 1 = \theta|X_1 = X_2) = \frac{1}{2} \times \frac{1}{2} = .25.$$

This $C(x_1, x_2)$ then has .75 coverage probability of smallest size.

In other words, conditional on $x_1 \neq x_2$ we are certain that $\bar{x} = \theta$ so $C(x_1, x_2)$ contains θ with 100% confidence but only with 50% confidence when $x_1 = x_2$. So it seems more sensible to state this after using the data rather than .75 before seeing the data and living by that. This brings to the fore the possible use of conditional confidence sets and the importance of principles of conditioning even when the Repeated Sample Principle is used.

A somewhat more intriguing conditioning example, purporting to solve the so-called ancient "Problem of the Nile," was introduced by Fisher (1956b) as follows:

> The agricultural land (x, y; geographical coordinates) of a pre-dynastic Egyptian village is of unequal fertility (probability density). Given the height (θ) which the Nile will rise, the fertility ($f(x, y|\theta)$) of every portion of it is known with exactitude, but the height of the flood affects different parts of the territory unequally. It is required to divide the area among the several households of the village so that the yields of the lots assigned to each shall be in predetermined proportions, whatever may be the height to which the river rises.

In summary, we have:

geographic location $= (x, y)$,
height of the Nile $= \theta$,
fertility $=$ probability density eg.

$$f(x, y|\theta) = e^{-\theta x - y/\theta},$$

proportionate yield of a lot $=$ probability of a region eg.

$$A = \int_A \int e^{-\theta x - \theta^{-1} y} dx dy.$$

The boundary of a lot should not depend on θ and regions should be in predetermined proportions (i.e., find regions A that are independent of θ).

Since

$$f(x, y|\theta) = e^{-\theta x - \theta^{-1} y},$$

$2\theta X$ and $2\theta^{-1} Y$ are independent and each χ_2^2, a chi-square variate with 2 degrees of freedom, and

$$\frac{2\theta^{-1} Y}{2\theta X} \sim \frac{T^2}{\theta^2} \sim F_{2,2},$$

where $F_{2,2}$ represents an F variate with 2 and 2 degrees of freedom. Consider $T = \sqrt{Y/X}$ and $U = \sqrt{XY}$. Then $x = u/t$, $y = tu$, and

$$\left|\frac{\partial(y, x)}{\partial(t, u)}\right| = \begin{vmatrix} u & t \\ -u & 1 \\ \frac{}{t^2} & t \end{vmatrix} = \frac{u}{t} + \frac{u}{t} = 2x.$$

So

$$f(u, t|\theta) = \frac{2u}{t} e^{-\theta\frac{u}{t} - \frac{tu}{\theta}} = \frac{2u}{t} e^{-u\left(\frac{\theta}{t} + \frac{t}{\theta}\right)},$$

$$f(t|\theta) = 2 \int_0^\infty \frac{u}{t} e^{-u\left(\frac{\theta}{t} + \frac{t}{\theta}\right)} du.$$

Let $u(\theta/t + t/\theta) = z$. Then

$$f(t|\theta) = \frac{2}{t} \int_0^\infty \frac{ze^{-z}}{\left(\frac{\theta}{t} + \frac{t}{\theta}\right)^2} dz = \frac{2}{t}\left(\frac{t}{\theta} + \frac{\theta}{t}\right)^{-2}.$$

Further,

$$f(u|\theta) = 2u \int \frac{1}{t} e^{-u\left(\frac{\theta}{t} + \frac{t}{\theta}\right)} dt.$$

Set $\log(t/\theta) = v$, $dt/t = dv$, thus

$$f(u|\theta) = 2u \int_{-\infty}^\infty e^{-u(e^{-v} + e^v)} dv = 2u \int_{-\infty}^\infty e^{-2u \cosh v} dv.$$

Now

$$K_m(z) = \int_0^\infty e^{-z \cosh v} \cosh(mv) dv$$

is Bessel's function of the second kind (or modified) and

$$K_0(2u) = \int_0^\infty e^{-2u \cosh v} dv,$$

where

$$\frac{1}{2}(e^{-v} + e^v) = \cosh v,$$

so that

$$f(u|\theta) = 2u \int_{-\infty}^{\infty} e^{-2u\cosh v} dv = 4u \int_{0}^{\infty} e^{-2u\cosh v} dv = 4uK_0(2u),$$

independent of θ. Hence

$$f(t|u, \theta) = \frac{f(t, u|\theta)}{f(u)} = \frac{\frac{2u}{t}e^{-u\left(\frac{\theta}{t} + \frac{t}{\theta}\right)}}{4uK_0(2u)}.$$

Now

$$\int_A \int e^{-\theta x - \theta^{-1}y} dx dy = \int_{A^*} \int f(t|u, \theta) f(u) dt du.$$

If t is integrated between 0 and infinity (since any other limits for t would involve θ) and u between any two constants, say u_1 and u_2, then

$$\int_{A^*} f(u) \, du = \int_{u_1}^{u_2} f(u) du$$

results in proportional yields independent of the height θ which is Fisher's solution to the "problem of the Nile."

Now this is really an introduction to the following problem. Consider again

$$f(x, y|\theta) = e^{-\theta x - y/\theta},$$
$$\log f = -\theta x - y/\theta,$$
$$\frac{d \log f}{d\theta} = -x + \frac{y}{\theta^2} = 0 \quad \text{or} \quad \theta^2 = \frac{y}{x}.$$

Hence the MLE $\hat{\theta} = \sqrt{Y/X} = T$ but $\hat{\theta}$ is not sufficient since the sufficient statistic is (X, Y). However, Fisher declares $U = \sqrt{YX}$ to be the ancillary statistic and suggests that inferences be made not by using $f(t|\theta)$ but on $f(t|u, \theta)$.

More generally, suppose $(X_1, Y_1), \dots, (X_n, Y_n)$ is a random sample so that

$$L(\theta) = e^{-(\theta \Sigma x_i + \theta^{-1} \Sigma y_i)},$$

then the MLE $\hat{\theta} = T = \sqrt{\Sigma Y_i / \Sigma X_i}$ but the sufficient statistic is $(\Sigma X_i, \Sigma Y_i)$. The information in the sample is easily calculated to be

$$I_n(\theta) = -E\left(\frac{d^2 \log L}{d\theta^2}\right) = \frac{2n}{\theta^2}.$$

Further letting, $U = \sqrt{(\Sigma X_i)(\Sigma Y_i)}$ then

$$f(u, t|\theta) = \frac{2e^{-u\left(\frac{t}{\theta} + \frac{\theta}{t}\right)} u^{2n-1}}{\Gamma^2(n)t},$$

$$f(t|\theta) = \frac{2\Gamma(2n)}{\Gamma^2(n)} \frac{1}{t} \left(\frac{t}{\theta} + \frac{\theta}{t}\right)^{-2n},$$

$$I_T(\theta) = -E\left[\frac{d^2 \log f_T}{d\theta^2}\right] = \frac{2n}{\theta^2} \cdot \frac{2n}{2n+1}.$$

So T is not fully informative.
Now

$$f(t|\theta, u) = \frac{f(t, u|\theta)}{f(u|\theta)},$$

and

$$f(u|\theta) = 4K_0(2u)\frac{u^{2n-1}}{\Gamma^2(n)}$$

is independent of θ. Thus U is an ancillary statistic. And

$$I_{T|U}(\theta) = \frac{2}{\theta^2} \frac{uK_1(2u)}{K_0(2u)}, \quad E_U I_{T|U}(\theta) = \frac{2n}{\theta^2},$$

so on averaging over U the information in the sample is recoverable. The following results can be obtained for the expectation, mean squared error (mse) and variance:

$$E(T|\theta) = \theta\frac{\Gamma(n+\frac{1}{2})\Gamma(n-\frac{1}{2})}{\Gamma^2(n)}, \quad E(T|U, \theta) = \theta\frac{K_1(2u)}{K_0(2u)},$$

$$\text{mse}(T) = E_U[\text{mse}(T|U)], \quad \text{var}(T) > E_U \text{ var}(T|U).$$

Now

$$f_{T|U}(t|u, \theta) = \frac{e^{-u\left(\frac{t}{\theta} + \frac{\theta}{t}\right)}}{2tK_0(2u)},$$

and

$$f_{V|U}(v|u) = \frac{e^{-2u\cosh v}}{2K_0(2u)},$$

where $V = \log\frac{T}{\theta}$ so that the distribution of V is independent of θ and is a pivot. Inference about θ is conditional on u and comes from either $f_{T|U}(t|u, \theta)$ or $f_{V|U}(v|u)$.

However in certain problems there may be more than one obvious ancillary. Here it was obvious. In non-obvious cases which ancillary should be used? Consider the following problem posed by Basu (1964):

Example 8.3

Let

$$L = f(n_1, n_2, n_3, n_4 | \theta) = \frac{n!}{n_1! n_2! n_3! n_4!} p_1^{n_1} p_2^{n_2} p_3^{n_3} p_4^{n_4}, \quad n = \sum_1^4 n_i$$

$$p_1 = \frac{1}{6}(1 - \theta), \quad p_2 = \frac{1}{6}(1 + \theta), \quad p_3 = \frac{1}{6}(2 - \theta), \quad p_4 = \frac{1}{6}(2 + \theta) \quad |\theta| \le 1.$$

Hence

$$\frac{d \log L}{d\theta} = -\frac{n_1}{1 - \theta} + \frac{n_2}{1 + \theta} - \frac{n_3}{2 - \theta} + \frac{n_4}{2 + \theta}.$$

The numerical sufficient statistic is (n_1, n_2, n_3, n_4) with one n_i being redundant. One can find the MLE $\hat{\theta}$ but $\hat{\theta}$ is not sufficient. Consider

$$U = (n_1 + n_2, n_3 + n_4) = (U_1, U_2) = (U_1, n - U_1),$$

and

$$f_{U_1}(u_1) = \binom{n}{u_1}(p_1 + p_2)^{u_1}(1 - p_1 - p_2)^{n - u_1} = \binom{n}{u_1}\left(\frac{1}{3}\right)^{u_1}\left(\frac{2}{3}\right)^{n - u_i}.$$

Consider also

$$V = (n_1 + n_4, n_2 + n_3) = (V_1, V_2) = (V_1, n - V_1)$$

and

$$f_{V_1}(v_1) = \binom{n}{v_1}\left(\frac{1}{2}\right)^n.$$

Therefore, both U and V are ancillary—so which one should we use?
Note

$$f(n_1, n_3 | U) = \binom{u_1}{n_1}\left(\frac{1 - \theta}{2}\right)^{n_1}\left(\frac{1 + \theta}{2}\right)^{u_1 - n_1}\binom{u_2}{n_3}\left(\frac{2 - \theta}{4}\right)^{n_3}\left(\frac{2 + \theta}{4}\right)^{u_2 - n_3},$$

$$f(n_1, n_3 | V) = \binom{v_1}{n_1}\left(\frac{1 - \theta}{3}\right)^{n_1}\left(\frac{2 + \theta}{3}\right)^{v_1 - n_1}\binom{v_2}{n_3}\left(\frac{2 - \theta}{3}\right)^{n_3}\left(\frac{1 + \theta}{3}\right)^{v_2 - n_3}.$$

Now for either conditional probability function we get the same solution to $d \log f/d\theta = 0$ and hence the same MLE, $\hat{\theta}$. In general then how do we discriminate between ancillaries. It has been suggested that one should prefer the ancillary that provides the most separation, that is, separating the relatively informative samples from the relatively uninformative ones. Let $T = (n_1, n_3)$ and denote the conditional information quantities as

$$I_{T|U}(\theta) = W(\theta) \quad \text{and} \quad I_{T|V}(\theta) = Z(\theta)$$

Note that

$$E_U[I_{T|U}(\theta)] = E_V[I_{T|V}(\theta)],$$

where

$$W(\theta) = \frac{\dfrac{3u_1}{n} + (1 - \theta^2)}{n(1 - \theta^2)(4 - \theta^2)}, \qquad Z(\theta) = \frac{\dfrac{2v_2\theta}{n} + (1 - \theta)(2 + \theta)}{n(1 - \theta^2)(4 - \theta^2)}.$$

So we may choose the ancillary that is more dispersed, that is, by calculating the variances of the conditional information, namely:

$$\frac{2}{n^3(1 - \theta^2)^2(4 - \theta^2)^2} = \text{var}(W(\theta)) > \text{var}(Z(\theta)) = \frac{\theta^2}{n^3(1 - \theta^2)^2(4 - \theta^2)^2}.$$

On this basis we would choose U as the ancillary to condition on (Cox, 1971).

8.4 BAYESIAN INTERVALS (SETS)

Recall from Section 6.1 that the posterior probability function is given as

$$p(\theta|D) \propto L(\theta|D)g(\theta),$$

where $g(\theta)$ is the prior probability function of $\theta \in \Theta$.

In cases where $X_1, \ldots, X_N$ are i.i.d., and $f(x|\theta)$ is absolutely continuous and θ is continuous over its range then

$$L(\theta|D) = \prod_{i=1}^{N} f(x_i|\theta).$$

Further,

$$p(\theta|D) = \frac{g(\theta) \prod_{i=1}^{N} f(x_i|\theta)}{\int g(\theta) \prod_{i=1}^{N} f(x_i|\theta) d\theta},$$

provided it is a bonafide probability function. The latter will be the case if $g(\theta)$ is a proper density and sometimes when $g(\theta)$ is improper.

For example, when $g(\theta)$ is uniform over an infinite segment of the real line, and therefore improper, in certain cases $p(\theta|D)$ still will be a proper probability function. For example, if $X_1, \ldots, X_N$ are i.i.d. $N(\mu, \sigma^2)$ and σ^2 known, then a simple computation yields $\mu \sim N(\bar{x}, \frac{\sigma^2}{n})$ for $g(\theta)$ uniform over the real line. Any confidence interval derived in Example 8.1 will be the same as the derived Bayesian interval with $1 - \alpha$ the probability that μ is in the interval rather than $1 - \alpha$ being a confidence coefficient representing the relative frequency of θ being included in that random interval in an infinite number of repetitions of drawing N independent observations from the $N(\mu, \sigma^2)$ population.

However, theoretically, the Bayesian approach is much more flexible in allowing probability statements to be calculated either analytically or numerically to obtain the shortest interval for a given probability or the maximum probability for an interval of a prescribed length. More generally, any probability statement regarding θ can be calculated, for instance, the probability that θ is between any values a and b or in some disconnected set of intervals, and so on. This enormous advantage, in general, requires a price, namely the assumption that one can either objectively or subjectively meaningfully define $g(\theta)$.

Subjectivists will attempt to elicit from an experimenter opinions concerning the chances of θ being in various intervals and thereby try to construct a plausible probability representation of a subjective $g(\theta)$. In the view of others the purpose of $g(\theta)$ is to represent minimal knowledge concerning θ so that its contribution to the posterior is dwarfed by the $L(\theta|D)$. This usually involves some reference prior generated by a method that involves invariance considerations as previously described in Section 6.3.

8.5 HIGHEST PROBABILITY DENSITY (HPD) INTERVALS

Suppose we have derived a posterior probability function for parameter θ as $p(\theta|x^{(N)})$. Using a particular loss function we can obtain the shortest size set of intervals for any given probability q, $0 \le q \le 1$.

Let the loss be defined for any set A of measurable intervals for θ be

$$L(A, \theta) = \begin{cases} \beta \mathcal{L}(A) - K & \text{if } \theta \in A \\ \beta \mathcal{L}(A) & \text{if } \theta \in A^c \end{cases}$$

for $K > 0$ where $\mathcal{L}(A)$ is Lebesque measure. Now

$$
\begin{aligned}
L(A) = E[L(A, \theta)] &= \int_A [\beta\mathcal{L}(A) - K]p(\theta|x^{(N)})d\theta + \int_{A^c} \beta\mathcal{L}(A)p(\theta|x^{(N)})d\theta \\
&= \int_{A \cup A^c} \beta\mathcal{L}(A)p(\theta|x^{(N)})d\theta - K \int_A p(\theta|x^{(N)})d\theta \\
&= \beta\mathcal{L}(A) - K \int_A p(\theta|x^{(N)})d\theta = \beta \int_A d\theta - K \int_A p(\theta|x^{(N)})d\theta \\
&= \int_A (\beta - Kp(\theta|x^{(N)}))d\theta.
\end{aligned}
$$

Hence

$$
\text{Inf}_A \ L(A) = L(A^*),
$$

where

$$
A^* = \left\{ \theta : p(\theta|x^{(N)}) \geq \frac{\beta}{K} \right\},
$$

the HPD region. While this is of comparative interest with regard to confidence intervals a Bayesian can always report the entire probability function.

Example 8.4

Suppose $X_1, \ldots, X_n$ are i.i.d. $f(x|\theta) = \theta e^{-\theta x}$, $\theta > 0$, $x \geq 0$. Then

$$
L(\theta|D) = \theta^n e^{-\theta t},
$$

where $t = \Sigma x_i$. Let $g(\theta) = \theta^{-1}$. Then this leads to

$$
p(\theta|t) = \frac{t^n \Theta^{n-1} e^{-n\Theta}}{\Gamma(n)},
$$

or $2\theta t \sim \chi^2_{2n}$ a chi-squared variate with $2n$ degrees of freedom but θ is the random variable and t is fixed. On the other hand, it is also clear that for θ fixed, $2\theta T \sim \chi^2_{2n}$ where T is the random variable. Hence confidence intervals have a Bayesian interpretation with the confidence coefficient being the posterior probability. We note that for $n = 1, p(\theta|t)$ is a decreasing function of θ. Therefore the HPD interval begins at $\theta = 0$ and stops at $\theta = \theta_0$, such that

$$
e^{-t\theta_0} = 1 - \alpha.
$$

For $n > 1$, $p(\theta|t) = 0$ at $\theta = 0$ and increases until the maximum $\theta = ((n-1)/t)$ and then decreases. Hence the HPD is a single interval calculated according to A^*. Again, it is to be emphasized the entire function $p(\theta|t)$ can be presented.

8.6 FIDUCIAL INFERENCE

R. A. Fisher proposed an eclectic inferential system that included at one level, like-lihood inference, and at another dependence on repeated sampling, sufficiency, and conditioning for estimation and testing. Included in this setup was a method for pro-ducing a "posterior" distribution without the benefit of a prior distribution under the rubric of a "fiducial argument." We now define and discuss Fisher's fiducial argument.

Let T be sufficient for $\theta \in (a, b)$, where both T and θ are scalars over the same range that is, $T \in (a, b)$. Further let $f_T(t|\theta)$ be a continuous density with distribution function $F_T(t|\theta)$. Let $1 - F(t|\theta)$ be continuous and monotonically increasing (or decreasing) in θ for each $t \in (a, b)$ with

(i) $\lim_{\theta \to a} (1 - F(t|\theta)) = 0$ and $\lim_{\theta \to b} (1 - F(t|\theta)) = 1$ if $1 - F$ is increasing

(ii) $\lim_{\theta \to a} F(t|\theta) = 0$ and $\lim_{\theta \to b} F(t|\theta) = 1$ if $1 - F$ is decreasing.

Then Fisher vests $1 - F(t|\theta) = \Phi(\theta|t)$ if (i) holds, or $F(t|\theta) = \Phi(\theta|t)$ if (ii) holds (since $\Phi(\theta|t)$ behaves exactly like a distribution function) with the property of being a posterior or "fiducial" distribution of θ for fixed t, claiming that the sufficient statistic T induces a distribution of values for θ. Note that in this case it is also a "confidence" distribution of values for θ.

The fiducial density of θ is then given by

$$\varphi(\theta|t) = \frac{d\Phi(\theta|t)}{d\theta} = \frac{d(1 - F(t|\theta))}{\theta} = -\frac{dF(t|\theta)}{d\theta} = \left|\frac{dF}{d\theta}\right|,$$

or

$$\varphi(\theta|t) = \frac{dF(t|\theta)}{d\theta} = \frac{dF}{d\theta}.$$

Therefore, fiducial probability statements about θ can be made for $\theta_0 \le \theta_1$,

$$P[\theta_0 \le \theta \le \theta_1] = \Phi(\theta_1|t) - \Phi(\theta_0|t).$$

In other words, the pivot $U = F(t|\theta)$ may be regarded as inducing a "fiducial" dis-tribution for θ for the observed value of t. The fiducial argument uses the obser-vations only to change the logical status of the parameter from one in which nothing is known of it, to the status of a random variable having a well-defined dis-tribution, according to Fisher.

Example 8.5

Assume $X_1, \ldots, X_n$ are i.i.d. $N(\theta, \sigma^2)$ where σ^2 is known. Now $\overline{X}$ is sufficient for θ where

$$f(\bar{x}|\theta) = \sqrt{\frac{n}{2\pi\sigma^2}} e^{-\frac{n(\bar{x}-\theta)^2}{2\sigma^2}}.$$

Further,

$$F(\bar{x}|\theta) = \int_{-\infty}^{\bar{x}} \sqrt{\frac{n}{2\pi\sigma^2}} e^{-\frac{n(u-\theta)^2}{2}} du = N\left(\frac{(\bar{x}-\theta)\sqrt{n}}{\sigma}\right) = \int_{-\infty}^{\frac{(\bar{x}-\theta)\sqrt{n}}{\sigma}} \frac{1}{\sqrt{2\pi}} e^{-\frac{y^2}{2}} dy,$$

where $N(\cdot)$ is the standard normal distribution function and

$$\varphi(\theta|\bar{x}) = \frac{-dF(\bar{x}|\theta)}{d\theta}$$

$$= \left|\left(\frac{-\sqrt{n}}{\sigma}\right) \frac{1}{\sqrt{2\pi}} e^{-\frac{(\bar{x}-\theta)^2 n}{2\sigma^2}}\right| = \sqrt{\frac{n}{2\pi\sigma^2}} e^{-\frac{n}{2\sigma^2}(\theta-\bar{x})^2},$$

such that, according to the fiducial argument, $\theta \sim N(\bar{x}, \sigma^2/n)$. Of course this is the equivalent of making $V = \frac{\bar{x}-\theta}{\sigma/\sqrt{n}}$ a pivot whose distribution is invariant if we consider $\bar{x}$ fixed and θ random. Note this is the same as the Bayesian posterior for a uniform prior on θ.

Example 8.6

Assume $X_1, \ldots, X_n$ are i.i.d. $f(x|\theta) = \theta e^{-\theta x}$, $\theta > 0$, $x \geq 0$. Then

$$L(\theta|D) = \theta^n e^{-\theta \sum_{i=1}^{n} x_i}, \quad T = \sum_{i=1}^{n} X_i \text{ is sufficient for } \theta.$$

Now $2\theta t = \chi_{2n}^2$ and

$$P(2\theta t < \chi_{2n}^2(1-\alpha)) = \int_0^{\chi_{2n}^2(1-\alpha)} f_{\chi_{2n}^2}(u) du = 1 - \alpha,$$

$$= P\left(\theta < \frac{\chi_{2n}^2(1-\alpha)}{2t}\right) = 1 - \alpha.$$

So θ has fiducial density corresponding to $\frac{\chi_{2n}^2}{2t}$,

$$\varphi(\theta|t) = \frac{te^{-t\theta}(t\theta)^{n-1}}{\Gamma(n)},$$

and we note this is the same probability function as the Bayesian posterior of Example 8.4.

8.7 RELATION BETWEEN FIDUCIAL AND BAYESIAN DISTRIBUTIONS

Since Fisher said the fiducial argument induced a distribution for θ and the only other mode of inference which posits a distribution for θ is the Bayesian mode, we may inquire whether these two modes of inference are related.

Recall

$$\varphi(\theta|t) = \frac{-dF(t|\theta)}{d\theta},$$

and the Bayesian approach when there is a single sufficient statistic yields

$$p(\theta|t) = \frac{g(\theta)f(t|\theta)h(x)}{\int g(\theta)f(t|\theta)h(x)d\theta} = \frac{g(\theta)f(t|\theta)}{\gamma(t)}$$

for prior probability function $g(\theta)$.

Now suppose we equate the fiducial and Bayesian posteriors,

$$\frac{-dF(t|\theta)}{d\theta} = \frac{g(\theta)f(t|\theta)}{\gamma(t)} = \frac{g(\theta)}{\gamma(t)}\frac{dF(t|\theta)}{dt}$$

or

$$\frac{-dF/d\theta}{dF/dt} = \frac{g(\theta)}{\gamma(t)}. \tag{8.7.1}$$

For a solution it is necessary that the left-hand side be equal to a product of a function of θ times a function of t. We rewrite (8.7.1)

$$\frac{1}{\gamma(t)}\frac{dF}{dt} + \frac{1}{g(\theta)}\frac{dF}{d\theta} = 0,$$

and set

$$\Gamma(t) = \int_{-\infty}^{t} \gamma(t')dt',$$

$$G(\theta) = \int_{-\infty}^{\theta} g(\theta')d\theta'$$

so

$$\frac{1}{\gamma}\frac{dF}{d\Gamma}\frac{d\Gamma}{dt} + \frac{1}{g}\frac{dF}{dG}\frac{dG}{d\theta} = 0.$$

For arbitrary $H(\Gamma, G)$, let

$$\frac{dF}{d\Gamma} + \frac{dF}{dG} = 0 = \frac{dH(\Gamma, G)}{d\Gamma} + \frac{dH(\Gamma, G)}{dG}.$$

Lindley (1958), who provided this argument, has shown from the theory of partial differential equations that the only solutions are $F = Const.$ which is not possible here or $F(t|\theta) = H[\Gamma(t) - G(\theta)]$ and H is arbitrary.

Now suppose $F = H(\Gamma(t) - G(\theta))$ and let $\Gamma(t) = u$ and $G(\theta) = \tau$ where u is a new random variable and τ a new parameter. Then for $F = H(u - \tau)$, τ is a location parameter for u, and F is decreasing in τ, and

$$\frac{-dH(u-\tau)}{d\tau} = \frac{-dH(u-\tau)}{d(u-\tau)}\frac{d(u-\tau)}{d\tau} = \frac{dH(u-\tau)}{d(u-\tau)}\frac{d(u-\tau)}{du} = \frac{dH(u-\tau)}{du},$$

or

$$\frac{\left(\dfrac{-dH(u-\tau)}{d\tau}\right)}{\dfrac{dH(u-\tau)}{du}} = \frac{\left(\dfrac{-dF}{d\theta}\dfrac{d\theta}{d\tau}\right)}{\dfrac{dF}{dt}\dfrac{dt}{du}} = 1,$$

or

$$\frac{-\left(\dfrac{dF}{d\theta}\right)}{\dfrac{dF}{dt}} = \frac{\left(\dfrac{dt}{du}\right)}{\dfrac{d\theta}{d\tau}} = \frac{\left(\dfrac{d\tau}{d\theta}\right)}{\dfrac{du}{dt}} = \frac{g(\theta)}{\gamma(t)}.$$

Hence,

$$\frac{-dF}{d\theta} = \frac{dF}{dt}\frac{g(\theta)}{\gamma(t)},$$

or

$$\varphi(\theta|t) = f(t|\theta)\frac{g(\theta)}{\gamma(t)},$$

as required and prior density $g_\tau(\tau) = $ constant since $G(\theta) = \tau$.

Hence, we showed that, a necessary and sufficient condition for a fiducial inversion to be equivalent to a Bayesian inversion is that, for some H

$$F(t|\theta) = H(u - \tau),$$
$$u = \Gamma(t), \quad \tau = G(\theta). \tag{8.7.2}$$

We note that if θ is a scale parameter for t then $\log \theta$ is a location parameter for $\log t$, that is, $u^* = \log t$ and $\tau^* = \log \theta$ with

$$F = H^*(\log t - \log \theta),$$

then

$$g(\tau^*)d\tau^* \propto d\tau^* \implies d \log \theta \propto \frac{1}{\theta} d\theta.$$

We now show that the use of the square root of $I(\theta)$ as a prior density is necessary for fiducial inversion. Since for equivalence $F(t|\theta) = H(u - \tau)$, then

$$f(t|\theta) = \frac{dF}{dt} = \frac{dH(u - \tau)}{d(u - \tau)} \times \frac{d(\Gamma(t) - G(\theta))}{dt},$$

$$\log f(t|\theta) = \log\left(\frac{dH(u - \tau)}{d(u - \tau)}\right) + \log\left(\frac{d\Gamma(t)}{dt}\right),$$

$$\frac{d \log f(t|\theta)}{d\theta} = \frac{d}{d\theta} \log\left(\frac{dH(u - \tau)}{d(u - \tau)}\right) \equiv \frac{d \log h(u - \tau)}{d\theta}$$

$$= \frac{d \log h(u - \tau)}{d(u - \tau)} \times \frac{d(u - \tau)}{d\theta}$$

$$= \frac{d \log h(u - \tau)}{d(u - \tau)} \left(\frac{-d\tau}{d\theta}\right).$$

So

$$E_T\left[\frac{d \log f(t|\theta)}{d\theta}\right]^2 = \left(\frac{d\tau}{d\theta}\right)^2 E_U\left(\frac{d \log h(u - \tau)}{d(u - \tau)}\right)^2.$$

Since $\frac{d\tau}{d\theta} = \frac{dG(\theta)}{d\theta}$,

$$I(\theta) = [g(\theta)]^2 E_V\left(\frac{d \log h(v)}{dv}\right)^2.$$

The second factor on the right-hand side above is independent of θ since $v = u - \tau$. Therefore, $I(\theta) \propto (g(\theta))^2$ and

$$I^{\frac{1}{2}}(\theta) \propto g(\theta).$$

Hence, the equivalence

$$F(t|\theta) = H(u - \tau) = H(\Gamma(t) - G(\theta))$$

yields $I(\theta) = (d\tau/d\theta)^2$, and $I^{\frac{1}{2}}(\theta)$ yields the Bayesian correspondent to the fiducial inversion. Thus the condition is necessary for fiducial and Bayesian procedures to yield equivalent results.

Fisher also claimed that once a fiducial distribution had been obtained it could be used as a prior distribution for a second sample just as in the Bayesian case. Recall for t a single sufficient statistic for θ

$$g(\theta)f(t|\theta) \propto p(\theta|t)$$

and for a second independent sample,

$$g(\theta)f(t_1|\theta)f(t_2|\theta) \propto p(\theta|t_1)f(t_2|\theta) \propto p(\theta|t_1, t_2),$$

where t_1 and t_2 represent independent samples.
Further,

$$p(\theta|t_1, t_2) \propto g(\theta)L(\theta|t_1)L(\theta|t_2) \propto g(\theta)f(t_1|\theta)f(t_2|\theta).$$

Now suppose from the fiducial inversion we get $\varphi(\theta|t_1)$ and we have an independent new sample $f(t_2|\theta)$. Then according to Fisher, the fiducial distribution based on the combined sample can be obtained as

$$\Psi(\theta|t_1, \ t_2) = \frac{\varphi(\theta|t_1)f(t_2|\theta)}{\int \varphi f d\theta},$$

which equals the posterior $p(\theta|t_1, t_2)$ provided (8.7.1) is satisfied in t_1.
If we had both samples at once and we inverted them so that

$$L(\theta|t_1, \ t_2) = L(\theta|t_1)L(\theta|t_2) \propto f(t|\theta),$$

where t is the total sufficient statistic that includes both samples $t = t(t_1, \ t_2)$ then Fisher asserts that

$$\varphi(\theta|t) = \frac{-dF(t|\theta)}{d(\theta)} = \Psi(\theta|t_1, \ t_2) \tag{8.7.3}$$

but without proof. Lindley (1957) gave the following argument.

The most general family that admits a single sufficient statistic for scalar θ is the exponential,

$$f(x|\theta) = e^{k(x)P(\theta)+Q(\theta)+C(x)}$$

$$= e^{k(x)P(\theta)}q(\theta)c(x).$$

It is clear that for $X_1, \ldots, X_n$ a random sample from the exponential family, $\sum_i k(X_i) = T$ is sufficient. To simplify, without loss of generality, we consider $T_1 = k(X_1)$ and $T_2 = k(X_2)$ so

$$L(\theta|t_1, t_2) \propto q^2(\theta)e^{P(\theta)(t_1+t_2)}.$$

The fiducial inversion using $t = t_1 + t_2$ is $\varphi(\theta|t)$ that is, a function of $t_1 + t_2$. Now consider the step by step fiducial inversion

$$\Psi(\theta|t_1, t_2) = \frac{\varphi(\theta|t_1)f(t_2|\theta)}{\int \varphi f d\theta}.$$

If Ψ is the same as φ it must be a function of $t_1 + t_2$. Now for θ versus θ'

$$\frac{\Psi(\theta|t_1, t_2)}{\Psi(\theta'|t_1, t_2)} = \frac{\varphi(\theta|t_1)f(t_2|\theta)}{\varphi(\theta'|t_1)f(t_2|\theta')}$$

should be a function of $t_1 + t_2$ when $\theta' \neq \theta$ and independent of the interchange of t_1 and t_2 so that

$$\frac{\varphi(\theta|t_1)f(t_2|\theta)}{\varphi(\theta'|t_1)f(t_2|\theta')} = \frac{\varphi(\theta|t_2)f(t_1|\theta)}{\varphi(\theta'|t_2)f(t_1|\theta')},$$

$$\frac{\varphi(\theta|t_1)\varphi(\theta'|t_2)}{\varphi(\theta'|t_1)\varphi(\theta|t_2)} = \frac{f(t_2|\theta')f(t_1|\theta)}{f(t_2|\theta)f(t_1|\theta')} = \frac{e^{t_2 P(\theta')+t_1 P(\theta)}}{e^{t_2 P(\theta)+t_1 P(\theta')}}$$

$$= e^{t_1[P(\theta)-P(\theta')]-t_2[P(\theta)-P(\theta')]}$$

$$= e^{(t_1-t_2)(P(\theta)-P(\theta'))}$$

or

$$\varphi(\theta|t_1) = \frac{\varphi(\theta'|t_1)e^{-t_1 P(\theta')}\varphi(\theta|t_2)e^{-t_2 P(\theta)}}{\varphi(\theta'|t_2)e^{-t_2 P(\theta')}}e^{t_1 P(\theta)}.$$

Regard θ' and t_2 as constants then

$$\varphi(\theta|t_1) = A(t_1)B(\theta)e^{t_1 P(\theta)}.$$

Hence,

$$\frac{-dF(t_1|\theta)/d\theta}{dF(t_1|\theta)/dt_1} = \frac{\varphi(\theta|t_1)}{f(t_1|\theta)} = \frac{A(t_1)B(\theta)e^{t_1 P(\theta)}}{h(t_1)q(\theta)e^{t_1 P(\theta)}}$$

$$\equiv \frac{g^*(\theta)}{\gamma^*(t_1)},$$

which implies equivalence of the Bayesian and Fiducial inversions based on t_1. Hence (8.7.2) $(F\ (t_1|\theta) = H(\Gamma(t_1) - G(\theta)))$ is necessary in t_1 for

$$\varphi(\theta|t_1, t_2) = \Psi(\theta|t_1, t_2)$$

to be true.

Moreover since the distribution of $t_1 + t_2$ is of exponential form it follows that (8.7.2) must also be satisfied in $t_1 + t_2$ since it is satisfied in t_1, thus (8.7.3) must be satisfied.

We now seek the relationship between the exponential family and $F(t|\theta) = H(\Gamma(t) - G(\theta))$ to show that it is not vacuous. First we give an example of a distribution from the exponential class for which two observations x_1 and x_2 demonstrate that

$$\varphi(\theta|x_1, x_2) \neq \Psi(\theta|x_1, x_2) = \frac{\varphi(\theta|x_1)f(x_2|\theta)}{\int \varphi f d\theta}.$$

Example 8.6

Lindley further presents the following counter example. Let

$$f(x|\theta) = \frac{\theta^2}{1+\theta}(1+x)e^{-x\theta}, \quad \text{for } x > 0, \quad \theta > 0$$

$$F(x|\theta) = 1 - e^{-x\theta} - \frac{x\theta e^{-x\theta}}{1+\theta}.$$

This belongs to the exponential class. Note $X_1 + X_2 = Z$ is sufficient for θ. Lindley shows that

$$\varphi(\theta|z) \neq \Psi(\theta|x_1, x_2) \neq \Psi(\theta|x_2, x_1),$$

which shows a lack of consistency within the fiducial argument if it is used in the same manner as the Bayesian argument.

In fact, as must be the case here, even for a single observation the fiducial inversion is not equivalent to a Bayesian inversion. In order for this to be shown we require a preliminary result known as Cauchy's equation for the general solution to the functional equation

$$h(u + \tau) = g(u) + k(\tau),$$

which is

$$g(u) = au + b_1, \quad k(\tau) = a\tau + b_2.$$

Proof: Set $\tau = 0$ in $h(u) = g(u) + k(0) = g(u) + c_1$ or $g(u) = h(u) - c_1$, and set $u = 0$ in $h(\tau) = g(0) + k(\tau) = c_2 + k(\tau)$ or $k(\tau) = h(\tau) - c_2$, such that $h(u + \tau) = h(u) + h(\tau) - c_1 - c_2$.

Now

$$\underbrace{h(u + \tau) - c_1 - c_2}_{f(u + \tau)} = \underbrace{h(u) - c_1 - c_2}_{f(u)} + \underbrace{h(\tau) - c_1 - c_2}_{f(\tau)},$$

known as Cauchy's equation.

For an equation of this type

$$f(x_1 + x_2) = f(x_1) + f(x_2), \quad \text{set } x_1 = x_2 = x$$

or

$$f(2x) = 2f(x).$$

By induction, it can be shown that

$$f(x_1 + \cdots + x_n) = f(x_1) + \cdots + f(x_n).$$

Now let $x_1 = \cdots = x_n = x$ then

$$f(nx) = nf(x).$$

For m a positive integer let $x = \frac{m}{n}t$, then $f(mt) = nf(\frac{m}{n}t)$. Now $f(mt) = mf(t)$ so $nf(\frac{m}{n}t) = f(mt) = mf(t)$ or $f(\frac{m}{n}t) = \frac{m}{n}f(t)$, so since $r = \frac{m}{n}$ is a positive rational number, $f(rt) = r f(t)$ holds for all positive rational r. For $x = 0$, $f(n + 0) = f(n) + f(0)$, therefore, $f(0) = 0$ and

$$f(n \times 0) = 0 f(0), \quad \text{then } f(0) = 0 \text{ is good for } r = 0.$$

Let $f(x + y) = f(x) + f(y)$ and let $y = -x$ for $x > 0$ so that $y < 0$. Hence $f(0) = f(x) + f(-x)$ but $f(0) = 0$, so $-f(x) = f(-x)$. Therefore it is also true for negative r as well. Now in $f(rt) = rf(t)$ set $t = 1$ and $f(r) = rf(1) = ra$ now if true for all

rational r and f is continuous it is true for all real r so the solution is $f(x) = ax$. Now $h(u) = f(u) + c_1 + c_2$, therefore $h(u) = au + c_1 + c_2$ so that $g(u) = au + c_2$ and $h(\tau) = f(\tau) + c_1 + c_2$, therefore $h(\tau) = a\,\tau + c_1 + c_2$, so $k(\tau) = a\tau + c_1$, as required.
   Recall the exponential family

$$f(x|\theta) = q(\theta)c(x)e^{h(x)P(\theta)}.$$

Without loss of generality we can consider

$$f(x|\theta) = q(\theta)c(x)e^{\theta x},$$

so the class is as wide as the class of functions $c(x)$ where

$$\int c(x)e^{\theta x}dx = \frac{1}{q(\theta)}$$

exists for all θ in its range. Therefore we want to find under what conditions there exists $u = u(x)$ and $\tau = \tau(\theta)$ such that $f(u|\tau)$ is a function of $u - \tau$, or letting $\tau = -\tau$, a function of $u + \tau$. We transform to

$$f(u|\tau) = q(\theta(\tau))c(x(u))e^{\theta(\tau)x(u)}\frac{dx(u)}{du},$$

$$\log f = \log q(\theta(\tau)) + \log c(x(u)) + \theta(\tau)x(u) + \log\frac{dx(u)}{du},$$

thus

$$\frac{d^2\log f}{d\tau\,du} = \theta'(\tau)x'(u).$$

Then consider $f(u|\tau)$ as a function of $u + \tau$ since this is the condition that is necessary and sufficient for the fiducial inversion to be Bayesian. Then

$$\frac{d^2\log f(u+\tau)}{du\,d\tau} = \frac{d^2\log f(u+\tau)}{d(u+\tau)^2},$$

which must be either a constant or a function $u + \tau$. This is because

$$\frac{d\log f(u+\tau)}{du} = \frac{d\log f}{d(u+\tau)} \times \frac{d(u+\tau)}{du} = \frac{d\log f}{d(u+\tau)},$$

$$\frac{d^2\log f(u+\tau)}{d\tau\,du} = \frac{d\left(\dfrac{d\log f}{d(u+\tau)}\right)}{d(u+\tau)} \cdot \frac{d(u+\tau)}{d\tau} = \frac{d^2\log f}{d(u+\tau)^2}.$$

Therefore, $\theta'(\tau)x'(u)$ is a constant or a function of $u + \tau$ and $\log \theta'(\tau)x'(u) = \log \theta'(\tau) + \log x'(u)$ is also a function of $u + \tau$. Therefore, by Cauchy's equation

$$\log \theta'(\tau) = a\tau + b_1, \quad \log x'(u) = au + b_2,$$

where a, b_1 and b_2 are constants.

First suppose $a \neq 0$ so that

$$\theta'(\tau) = e^{a\tau + b_1}, \quad x'(u) = e^{au + b_2},$$

thus

$$\theta(\tau) = d_1 e^{a\tau} + c_1, \quad x(u) = d_2 e^{au} + c_2.$$

It is no loss of generality to let $c_1 = c_2 = 0$ since the terms involving them in

$$f(u|\tau) = q(\theta(\tau))c(x(u))e^{\theta(\tau)x(u)}\frac{dx}{du}$$

can be incorporated in the functions of u and τ as constants such that

$$f(u|\tau) = q_1(\tau)c_1(u)e^{d_1 d_2 e^{a(u+\tau)}}.$$

Therefore, $q_1(\tau)c_1(u)$ must be a function of $u + \tau$. Applying the same arguments as before,

$$q_1(\tau) = \beta_1 e^{a\tau} \quad \text{and} \quad c_1(u) = \beta_2 e^{au},$$

so

$$f(u|\tau) = \gamma e^{a(u+\tau)}e^{d_1 d_2 e^{a(u+\tau)}}, \quad \gamma = \beta_1\beta_2.$$

Transforming back to x, since $\theta x = d_1 d_2 e^{a(u+\tau)}$ or $\theta = d_1 e^{a\tau}$, then

$$\theta dx = d_1 d_2 a e^{a(u+\tau)} du, \quad dx = d_2 a e^{au} du = ax du \quad \text{and} \quad du = \frac{dx}{ax}.$$

Hence, for $\gamma = \beta_1\beta_2$

$$\left(\frac{\theta x}{d_1 d_2}\right)^{\alpha/a} = e^{\alpha(u+\tau)},$$

and

$$f(x|\theta) = \frac{\gamma x^{-1}}{a} \left(\frac{\theta x}{d_1 d_2}\right)^{\alpha/a} e^{\theta x}.$$

Since we assumed the range of x is independent of θ the range of u is independent of τ for u and τ over the whole real line and x and θ over the half lines ending at 0. Let $\frac{\alpha}{a} = \lambda$. Then for $x > 0$, $\theta < 0$, but by changing the sign of θ we obtain

$$f(x|\theta) \propto x^{\lambda-1} e^{-x\theta},$$

the gamma density.

Now returning to

$$\log \theta'(\tau) = a\tau + b_1,$$
$$\log x'(u) = au + b_2,$$

we consider the case $a = 0$. Then with an obvious re-definition of the b's, $\theta'(\tau) = b_1$ and $x'(u) = b_2$ such that

$$\theta(\tau) = b_1\tau + d_1, \quad x(u) = b_2 u + d_2,$$

so

$$f(u|\tau) = q_1(\tau)c_1(u)e^{2\delta u\tau}.$$

We can complete the square by multiplying by $e^{u^2\delta} \cdot e^{\tau^2\delta}$ and dividing by the same quantity yields

$$f(u|\tau) = q_2(\tau)c_2(u)e^{\delta(u+\tau)^2},$$

such that $q_2(\tau)c_2(u)$ must be a function of $u + \tau$ or a constant. So

$$q_2(\tau)c_2(u) = \beta e^{\alpha(u+\tau)},$$

or a constant, that is, $\alpha = 0$. Then

$$f(u|\tau) = \gamma e^{\alpha(u+\tau)} e^{\delta(u+\tau)^2}.$$

Let $z = u + \tau$ then

$$f(u|\tau) = \gamma e^{\alpha z + \delta z^2} = \gamma e^{\delta(z^2 + \frac{\alpha}{\delta}z)}$$

$$= c_1 e^{\delta(z^2 + \frac{\alpha}{\delta}z + \frac{\alpha^2}{4\delta^2})}$$

$$= c_1 e^{\delta(z+a)^2} = c_1 e^{\delta(u+\tau+a)^2},$$

so that U is normally distributed.

Since X is a linear function of U then X is also normally distributed. Hence the only continuous distributions of the exponential family that satisfy the condition

$$F(t|\theta) = H(\Gamma(t) - G(\theta)),$$

which render a fiducial inversion equivalent to a Bayesian inversion are the gamma and normal forms or transformed thereto. Hence it is clear that when

$$F(t|\theta) = H(\Gamma(t) - G(\theta)),$$

we can make fiducial, Bayesian and confidence procedures coincide though we are restricted to the normal and gamma families.

8.8 SEVERAL PARAMETERS

Let

$$X_{11}, \ldots, X_{1n_1} \text{ i.i.d. } N(\mu_1, \sigma^2), \qquad X_{21}, \ldots, X_{2n_2} \text{ i.i.d. } N(\mu_2, \sigma^2).$$

For the confidence solution, where $(n_1 + n_2 - 2)s^2 = \sum_{i=1}^{2} \sum_{j=1}^{n_i} (X_{ij} - \overline{X}_i)^2$,

$$t = (\overline{X}_1 - \overline{X}_2 - (\mu_1 - \mu_2))/s\sqrt{\frac{1}{n_1} + \frac{1}{n_2}} \sim t_{n_1+n_2-2},$$

a student t variate with $n_1 + n_2 - 2$ degrees of freedom, independently of the parameters. Thus

$$P[-t_{\alpha/2} \le t \le t_{\alpha/2}] = 1 - \alpha \quad \text{yields for } \delta = \mu_1 - \mu_2,$$

$$P\left[\overline{X}_1 - \overline{X}_2 - t_{\alpha/2}s\sqrt{\frac{1}{n_1} + \frac{1}{n_2}} \le \delta \le \overline{X}_1 - \overline{X}_2 + t_{\alpha/2}s\sqrt{\frac{1}{n_1} + \frac{1}{n_2}}\right] = 1 - \alpha.$$

To apply the fiducial argument we first reparametrize the likelihood

$$L(\mu_1, \mu_2, \sigma^2) = L(\delta, \mu_w, \sigma^2),$$

where

$$\mu_w = \frac{n_1\mu_1 + n_2\mu_2}{n_1 + n_2} \quad \text{and} \quad \bar{x}_w = \frac{n_1\bar{x}_1 + n_2\bar{x}_2}{n_1 + n_2}.$$

Then

$$L(\mu_1, \mu_2, \sigma^2) = L(\delta, \mu_w, \sigma^2)$$

$$= \frac{e^{-\frac{1}{2\sigma^2}[(n_1+n_2-2)s^2+(n_1+n_2)(\bar{x}_w-\mu_w)^2+\frac{n_1 n_2}{n_1+n_2}(\bar{x}_1-\bar{x}_2-\delta)^2]}}{\sigma^{n_1+n_2}},$$

or

$$L(\delta, \mu_w, \sigma^2) = L(\delta, \sigma^2)L(\sigma^2, \mu_w).$$

Now Fisher gives two arguments for a fiducial distribution on δ. First let

$$t = \frac{\bar{x}_1 - \bar{x}_2 - \delta}{s\sqrt{\frac{1}{n_1} + \frac{1}{n_2}}}$$

be the pivot from which

$$\varphi(t) \propto \left(1 + \frac{t^2}{n_1 + n_2 - 2}\right)^{-\frac{n_1+n_2-1}{2}}.$$

Then transform to the random variable δ

$$\varphi(\delta) \propto \left(1 + \frac{(\delta - (\bar{x}_1 - \bar{x}_2))^2}{\frac{s^2(n_1 + n_2)(n_1 + n_2 - 2)}{n_1 n_2}}\right)^{-\frac{n_1+n_2-1}{2}}.$$

Another argument he uses is a step-by-step approach. Let $\bar{x}_1 - \bar{x}_2 = \bar{d}$ and $v = n_1 + n_2 - 2$. Then

$$f(\bar{d}, \bar{x}_w, s^2) = n\left(\bar{d}\,\middle|\,\delta, \sigma^2\left(\frac{1}{n_1} + \frac{1}{n_2}\right)\right)n\left(\bar{x}_w\,\middle|\,\mu_w, \frac{\sigma^2}{n_1 + n_2}\right) \times f(s^2|\sigma^2),$$

where $n(\cdot)$ is the normal density function. He first inverts $f(s^2|\sigma^2)$ and obtains

$$\varphi(\sigma^2|s^2) = \frac{1}{2^{\nu/2}\Gamma\left(\frac{\nu}{2}\right)}\left(\frac{\nu s^2}{\sigma^2}\right)^{\frac{\nu}{2}}\frac{e^{-\frac{\nu s^2}{2\sigma^2}}}{\sigma^2}.$$

Fisher then inverts

$$1 - F(\bar{d}|\delta, \sigma^2) = \Phi(\delta|\bar{d}, \sigma^2)$$

to obtain

$$\varphi(\delta|\bar{d}, \sigma^2) = \frac{1}{\sqrt{2\pi\sigma^2\left(\frac{1}{n_1}+\frac{1}{n_2}\right)}}e^{-\frac{(\delta-\bar{d})^2}{2\sigma^2\left(\frac{1}{n_1}+\frac{1}{n_2}\right)}}.$$

Similarly,

$$\varphi(\mu_w|\sigma^2) = n\left(\bar{x}_w, \frac{\sigma^2}{n_1+n_2}\right)$$

and thus

$$\varphi(\mu_w, \delta, \sigma^2) = \varphi(\mu_w, \delta|\sigma^2)\varphi(\sigma^2) = \varphi(\sigma^2)\varphi(\delta|\sigma^2)\varphi(\mu_w|\sigma^2).$$

Then

$$\int_{-\infty}^{\infty}\varphi(\mu_w|\sigma^2)d\mu_w\int_0^{\infty}\varphi(\delta, \sigma^2)d\sigma^2 = \int_0^{\infty}\varphi(\delta|\sigma^2)\varphi(\sigma^2)d\sigma^2$$

$$= \varphi(\delta|\bar{d}, s^2) \propto \left(1+\frac{(\delta-\bar{d})^2n_1n_2}{s^2(n_1+n_2)(n_1+n_2-2)}\right)^{-\frac{n_1+n_2-1}{2}}.$$

From the Bayesian standpoint we get the same result for prior density

$$g(\mu_1, \mu_2, \sigma^2) \propto \frac{1}{\sigma^2}.$$

So again all three modes can be made to yield the same interval on δ.

8.9   THE FISHER-BEHRENS PROBLEM

Let

$$X_{11}, \ldots, X_{1n_1} \quad \text{i.i.d.} \quad N(\mu_1, \sigma_1^2),$$

$$X_{21}, \ldots, X_{2n_2} \quad \text{i.i.d.} \quad N(\mu_2, \sigma_2^2).$$

Now the joint set of sufficient statistics for $(\mu_1, \mu_2, \sigma_1^2, \sigma_2^2)$ is $(\bar{X}_1, \bar{X}_2, s_1^2, s_2^2)$, where $(n_i - 1)s_i^2 = \sum_1^{n_i} (X_{ij} - \bar{X}_i)^2$ so that

$$f(\bar{x}_1, \bar{x}_2, s_1^2, s_2^2 | \mu_1, \mu_2, \sigma_1^2, \sigma_2^2) = \prod_{i=1}^{2} n\left(\bar{x}_i, \frac{\sigma_i^2}{n_i}\right) f_{\chi_{n_i-1}^2}\left(\frac{(n_i-1)s_i^2}{\sigma_i^2}\right).$$

As before, transforming μ_1 and μ_2 to δ and μ_w, and inverting the two $\chi_{n_i-1}^2$ distributions, we obtain the fiducial density

$$\varphi(\delta, \mu_w, \sigma_1^2, \sigma_2^2) = \varphi(\delta | \sigma_1^2, \sigma_2^2)\varphi(\mu_w | \sigma_1^2, \sigma_2^2)\varphi(\sigma_1^2)\varphi(\sigma_2^2).$$

Integrating out μ_w and setting $Z = (\delta - \bar{d})/\left(\frac{s_1^2}{n_1} + \frac{s_2^2}{n_2}\right)^{\frac{1}{2}}$ results in

$$\varphi(z, \sigma_1^2, \sigma_2^2) = n\left(z \middle| 0, \left(\frac{\sigma_1^2}{n_1} + \frac{\sigma_2^2}{n_2}\right) \middle/ \left(\frac{s_1^2}{n_1} + \frac{s_2^2}{n_2}\right)\right)\varphi(\sigma_1^2)\varphi(\sigma_2^2),$$

and

$$\varphi(z) = \int_0^{\infty} \int_0^{\infty} \frac{e^{-\frac{z^2}{2}\left[\left(\frac{s_1^2}{n_1} + \frac{s_2^2}{n_2}\right) \middle/ \left(\frac{\sigma_1^2}{n_1} + \frac{\sigma_2^2}{n_2}\right)\right]} d\sigma_1^2 d\sigma_2^2}{\sqrt{2\pi\left(\frac{\sigma_1^2}{n_1} + \frac{\sigma_2^2}{n_2}\right) \middle/ \left(\frac{s_1^2}{n_1} + \frac{s_2^2}{n_2}\right)}\varphi(\sigma_1^2)\varphi(\sigma_2^2)}.$$

Note $\varphi(-z) = \varphi(z)$ so the marginal distribution of Z is symmetric about 0.
    Fisher also shows

$$\Phi(z) = \int_{-\infty}^{\infty} f(t_2)\left[\int_{-\infty}^{\frac{z}{\sin\theta} + \frac{t_2}{\tan\theta}} f(t_1)dt_1\right]dt_2,$$

where $t_i = \frac{\bar{x}_i - \mu_i}{s_i/\sqrt{n_i}}$, $i = 1, 2$,

$$\tan\theta = \frac{s_1/\sqrt{n_1}}{s_2/\sqrt{n_2}}, \quad \sin\theta = \frac{s_1/\sqrt{n_1}}{\left(\frac{s_1^2}{n_1} + \frac{s_2^2}{n_2}\right)^{\frac{1}{2}}}$$

and

$$f(t_i) = \frac{\Gamma\left(\frac{n_i}{2}\right)}{\sqrt{\pi(n_i - 1)}} \frac{\left(1 + \frac{t_i^2}{n_i - 1}\right)^{-\frac{n_i}{2}}}{\Gamma\left(\frac{n_i - 1}{2}\right)}.$$

This has been tabled by Fisher and one can obtain

$$P\left(-z_{\frac{\alpha}{2}} \le Z \le z_{\frac{\alpha}{2}}\right) = 1 - \alpha$$

or

$$P\left[\bar{d} - \left(\frac{s_1^2}{n_1} + \frac{s_2^2}{n_2}\right)^{\frac{1}{2}} z_{\frac{\alpha}{2}} \le \delta \le \bar{d} + \left(\frac{s_1^2}{n_1} + \frac{s_2^2}{n_2}\right)^{\frac{1}{2}} z_{\frac{\alpha}{2}}\right] = 1 - \alpha.$$

This solution is equivalent to a Bayes solution with prior density

$$g(\mu_1, \mu_2, \sigma_1^2, \sigma_2^2) \propto \frac{1}{\sigma_1^2 \sigma_2^2}.$$

We can find the explicit density for $n_1 = n_2 = 2$. Let

$$W = \frac{(\delta - \bar{d})\sqrt{2}}{s_1 + s_2} = \frac{Z(s_1^2 + s_2^2)^{\frac{1}{2}}}{s_1 + s_2}, \quad Z = \frac{\delta - \bar{d}}{\left(\frac{s_1^2}{2} + \frac{s_2^2}{2}\right)^{\frac{1}{2}}}.$$

Now

$$\varphi(w, \sigma_1^2, \sigma_2^2) = n\left(w | 0, \frac{\sigma_1^2 + \sigma_2^2}{(s_1 + s_2)^2}\right) \varphi(\sigma_1^2 | s_1^2) \varphi(\sigma_2^2 | s_2^2),$$

since

$$\frac{(\delta - \bar{d})\sqrt{2}}{s_1 + s_2}$$

has a normal fiducial density with mean 0 and variance

$$\frac{\sigma_1^2 + \sigma_2^2}{(s_1 + s_2)^2}$$

conditional on σ_1^2 and σ_2^2. Then

$$\varphi(\sigma_1^2, \sigma_2^2|s_1^2, s_2^2) \propto \frac{e^{-\frac{s_1^2}{2\sigma_1^2} - \frac{s_2^2}{2\sigma_2^2}}}{\sigma_1^3 \sigma_2^3}.$$

Let $\frac{\sigma_2^2}{\sigma_1^2} = \rho$, $\sigma_2^2 = \rho\sigma_1^2$, $d\sigma_2^2 = \sigma_1^2 d\rho$, then

$$\varphi(w, \sigma_1^2, \rho) \propto n\left(w|0, \frac{\sigma_1^2(1+\rho)}{(s_1+s_2)^2}\right) \frac{e^{-\frac{s_1^2}{2\sigma_1^2}} e^{-\frac{s_2^2}{2\rho\sigma_1^2}}}{\sigma_1^3} \frac{\sigma_1^2}{\rho^{\frac{3}{2}}\sigma_1^3}$$

$$\propto \frac{1}{\sigma_1\sqrt{1+\rho}} e^{-\frac{w^2}{2}\frac{(s_1+s_2)^2}{\sigma_1^2(1+\rho)}} \frac{e^{-\frac{1}{2\sigma_1^2}(s_1^2+\frac{s_2^2}{\rho})}}{\sigma_1^4 \rho^{\frac{3}{2}}}.$$

Hence,

$$\varphi(w, \rho) \propto \frac{1}{\rho^{\frac{3}{2}}\sqrt{1+\rho}} \int \frac{e^{-\frac{1}{2\sigma_1^2}\left(\frac{w^2(s_1+s_2)^2}{1+\rho} + s_1^2 + \frac{s_2^2}{\rho}\right)}}{\sigma_1^5} d\sigma_1^2.$$

Let $y = \frac{1}{\sigma_1^2}$, then

$$\varphi(w, \rho) \propto \frac{1}{\rho^{\frac{3}{2}}\sqrt{1+\rho}\left[\frac{w^2(s_1+s_2)^2}{1+\rho} + s_1^2 + \frac{s_2^2}{\rho}\right]^{\frac{3}{2}}}$$

and

$$\int_0^\infty \varphi(w, \rho) d\rho \propto \frac{1}{(s_1+s_2)^2 + w^2(s_1+s_2)^2} \propto \frac{1}{1+w^2},$$

or

$$\varphi(w) = \frac{1}{\pi(1+w^2)},$$

a Cauchy variable.

Fisher also obtains for $r = s_2^2/s_1^2$,

$$\varphi(w|\rho, r) = \frac{\varphi(w, \rho)}{\int \varphi(w, \rho)dw}$$

$$\propto \left(1 + \frac{w^2(1 + r^{\frac{1}{2}})^2}{(1 + \rho)\left(1 + \frac{r}{\rho}\right)}\right)^{-\frac{3}{2}} = f(w|\rho, r).$$

Now one can obtain the unconditional sampling distribution of w given ρ as

$$f(w|\rho) = \frac{1}{\pi(1 + w^2)}\left[\sqrt{\frac{\rho}{1 + \rho + w^2}} + \sqrt{\frac{1}{1 + \rho + \rho w^2}}\right]$$

(Geisser, 1969). Note that $f(w|\rho) = f(-w|\rho)$, and

$$\lim_{\rho \to 0} f(w|\rho) = \frac{1}{\pi(1 + w^2)} = \lim_{\rho \to \infty} f(w|\rho).$$

The distribution function is

$$F(w|\rho) = \frac{1}{2} + \frac{1}{\pi}\left\{\sin^{-1}\frac{w\sqrt{\rho}}{\sqrt{(1 + \rho)(1 + w^2)}} + \sin^{-1}\frac{w}{\sqrt{(1 + \rho)(1 + w^2)}}\right\}$$

and

$$P[W \geq w_\alpha|\rho] = 1 - \frac{1}{2} - \frac{1}{\pi}\left\{\sin^{-1}\frac{w_\alpha\sqrt{\rho}}{\sqrt{(1 + \rho)(1 + w_\alpha^2)}} + \sin^{-1}\frac{w_\alpha}{\sqrt{(1 + \rho)(1 + w_\alpha^2)}}\right\}$$

$$= \alpha(\rho).$$

For $w_\alpha > 0$,

$$P[-w_\alpha \leq W \leq w_\alpha] = 1 - 2\alpha(\rho),$$

$$P\left[\bar{d} - \left(\frac{s_1 + s_2}{\sqrt{2}}\right)w_\alpha \leq \delta \leq \bar{d} + \left(\frac{s_1 + s_2}{\sqrt{2}}\right)w_\alpha\right] = 1 - 2\alpha(\rho).$$

We now seek the value of ρ that maximizes $\alpha(\rho)$ and hence minimizes $1 - 2\alpha(\rho)$. It can easily be shown that for any fixed w_α, or a fixed length of the interval, the probability is minimized or, conversely, for a fixed probability the length of the interval is maximized when $\rho = 0$ or ∞. Note that this would be the fiducial interval for δ. Hence the most conservative limits for δ are given by the fiducial solutions where $\rho = 0$ or ∞, values that are not very appealing for $\rho = \sigma_2^2/\sigma_1^2$. Here we have a situation that the fiducial inversion which is equivalent to a Bayesian inversion will not coincide with a confidence solution, and when using the usual improper prior has an awkward feature.

8.10   CONFIDENCE SOLUTIONS

We now discuss two confidence solutions that have been proposed.

1. *Random Pairing, Scheffé (1943)*:
   It will be sufficient to indicate the case where $n_1 = n_2 = n$, say. It is suggested the observations be randomly paired or assumed that the numbering of the observations are independent of the values of the observations. Let

   $$Y_i = X_{1i} - X_{2i} \quad i = 1, \ldots, n.$$

Then Y_i are i.i.d. and $N(\mu_1 - \mu_2, \tau^2)$ such that

$$\overline{Y} \sim N\left(\delta, \frac{\tau^2}{n}\right), \quad (n-1)s^2 = \Sigma(Y_i - \overline{Y})^2, \quad n\overline{Y} = \sum_1^n Y_i.$$

Then

$$t = \frac{\overline{Y} - \delta}{s/\sqrt{n}} \sim t_{n-1},$$

a student variate with $n - 1$ degrees of freedom. Then

$$P\left[\overline{Y} - \frac{s}{\sqrt{n}}t_{\frac{\alpha}{2}} \le \delta \le \overline{Y} + \frac{s}{\sqrt{n}}t_{\frac{\alpha}{2}}\right] = 1 - \alpha.$$

Note that no matter what the random pairing is $\overline{Y} = \overline{X}_1 - \overline{X}_2$ but s^2 depends heavily on the random pairing. This is unlikely to be acceptable in statistical practice. In the case $n_1 < n_2$ the following trick has been proposed to keep $\overline{Y} = \overline{X}_1 - \overline{X}_2$. Let

$$Y_i = X_{1i} - \sqrt{\frac{n_1}{n_2}}X_{2i} + \frac{1}{\sqrt{n_1 n_2}}\sum_1^{n_1}X_{2i} - \overline{X}_2 \quad i = 1, \ldots, n_1,$$

which preserves $\overline{Y} = \overline{X}_1 - \overline{X}_2$. However, these solutions have never been given serious consideration, even by frequentists.

2. *Solution by Welch (1947):*
Consider the Behrens-Fisher statistic

$$Z = \frac{(\overline{X}_1 - \overline{X}_2) - (\mu_1 - \mu_2)}{\left(\dfrac{s_1^2}{n_1} + \dfrac{s_2^2}{n_2}\right)^{\frac{1}{2}}},$$

and let

$$c = \frac{s_1^2/n_1}{s_1^2/n_1 + s_2^2/n_2}, \qquad \gamma = \frac{\sigma_1^2/n_1}{\sigma_1^2/n_1 + \sigma_2^2/n_2}.$$

Now Welch considers

$$\int_0^\infty P[|Z| < V(c)|c, \gamma] f(c|\gamma) dc = F(\gamma)$$

$$= P[|Z| < V(c)|\gamma],$$

and it appears from numerical calculations that $F(\gamma)$ seems to be almost independent of γ. $V(c)$ is a rather complicated function of c and does not particularly concern us. The rationale for Welch's approach is the following: First consider the case where $\sigma_1^2 = \sigma_2^2$. Then for the "student" case suppose

$$P[\overline{d} - \delta > h(s^2, q)|\sigma^2] = q \quad \text{independent of } \sigma^2.$$

We can find such $h(s^2, q)$, namely,

$$h(s^2, q) = t_q s \sqrt{\frac{1}{n_1} + \frac{1}{n_2}}$$

for t_q defined as

$$q = \int_{t_q}^\infty s(t) dt,$$

where $s(t)$ represents the density of $t_{n_1+n_2-2}$. By analogy, Welch says, for $\sigma_1^2 \neq \sigma_2^2$,

$$P[\overline{d} - \delta > h(s_1^2, s_2^2, q)] = q$$

and if q is independent of all parameters for some function $h(\cdot)$ then we have a bonafide confidence interval.

Tables for this method appear in *Biometrika Tables for Statisticians* by Pearson and Hartley (1966). We extract the relevant heading first line of Table 11.

Table 11: (From Pearson and Hartley, 1966). Test for Comparisons Involving Two Variances Which Must be Separately Estimated. Upper 5% Critical Values of Z (i.e. Upper 10% Critical Values of |Z|)

c	$\dfrac{\lambda_1 s_1^2}{\lambda_1 s_1^2 + \lambda_2 s_2^2}$	0.0	0.1	0.2	0.3	0.4	0.5	0.6	0.7	0.8	0.9	1.0
v_2	v_1											
6	6	1.94	1.90	1.85	1.80	1.76	1.74	1.76	1.80	1.85	1.90	1.94
	8	1.94	1.90	1.85	1.80	1.76	1.73	1.74	1.76	1.79	1.82	1.86
	10	1.94	1.90	1.85	1.80	1.76	1.73	1.73	1.74	1.76	1.78	1.81
	15	1.94	1.90	1.85	1.80	1.76	1.73	1.71	1.71	1.72	1.73	1.75
	20	1.94	1.90	1.85	1.80	1.76	1.73	1.71	1.70	1.70	1.71	1.72
	∞	1.94	1.90	1.85	1.80	1.76	1.72	1.69	1.67	1.66	1.65	1.64

We now review the problem from Fisher's point of view. The relevant entities for the two populations Π_1, and Π_2 are:

data	Π_1 is $N(\mu_1, \sigma_1^2)$ $n_1, \bar{x}_1, s_1^2$		Π_2 is $N(\mu_2, \sigma_2^2)$ $n_2, \bar{x}_2, s_2^2$
pivotal		$Z = \dfrac{(\bar{x}_1 - \bar{x}_2) - (\mu_1 - \mu_2)}{\sqrt{s_1^2/n_1 + s_2^2/n_2}}$	
weights	$c = \dfrac{s_1^2/n_1}{s_1^2/n_1 + s_2^2/n_2}$		$\gamma = \dfrac{\sigma_1^2/n_1}{\sigma_1^2/n_1 + \sigma_2^2/n_2}$

Note that when $n_1 = n_2$, Z is equivalent to the t statistic and if in addition $\gamma = 0.5$, $\sigma_1^2 = \sigma_2^2$ then $P[|t| < 1.782] = 0.9$. Here $P[|Z| < 1.74] = 0.9$. This is then not an acceptable procedure according to Fisher (1956a).

Now $\int_0^\infty P[|Z| < V(c)|c, \gamma]f(c|\gamma)dc = F(\gamma) = 0.90$ no matter what γ is, at least to 3 decimal places.

Fisher notes that if $c = 0.5$ and $V(c) = 1.74$ no value in Table 12 reaches the stated probability 0.9 no matter what γ is. Welch (1956) retorts that for any fixed value of γ, the average over all values of c is 0.9 to 3 decimal places. In a sense this tells us what is going on. Welch considers

$$\int_0^\infty P[|Z| < V(c)|c, \gamma]f(c|\gamma)dc = P[|Z| < V(c)|\gamma],$$

Table 12: (From Pearson and Hartley, 1966). Tabular Values of $P[|Z| < V(c)|c, \gamma]$ for Sample Size $n_1 = n_2 = 7$, Modified from Welch (1956). For $\nu_1 = \nu_2 = 6$, $\lambda_1 = \lambda_2 = \frac{1}{7}$, $\frac{\lambda_1 s_1^2}{\lambda_1 s_1^2 + \lambda_2 s_2^2} = \frac{s_1^2}{s_1^2 + s_2^2} = c$.

		γ								
c	V(c)	.1	.2	.3	.4	.5	.6	.7	.8	.9
.0	1.94	.976	.970	.960	.944	.924	.892	.842	.756	.598
.1	1.90	.918	.944	.944	.934	.918	.890	.844	.764	.608
.2	1.85	.850	.910	.924	.922	.910	.888	.846	.772	.620
.3	1.80	.784	.876	.902	.908	.902	.884	.848	.780	.634
.4	1.76	.728	.840	.882	.896	.896	.884	.852	.792	.654
.5	1.74	.682	.810	.862	.886	.892	.886	.862	.810	.682
.6	1.76	.654	.792	.852	.884	.896	.896	.882	.840	.728
.7	1.80	.634	.780	.848	.884	.902	.908	.902	.876	784
.8	1.85	.620	.772	.846	.888	.910	.922	.924	.910	.850
.9	1.90	.608	.764	.844	.890	.918	.934	.944	.944	.918
1.0	1.94	.598	.756	.842	.892	.924	.944	.960	.970	.976

which appears to be approximately independent of γ. Fisher considers

$$\int_0^\infty P[|Z| < V(c)|c, \gamma]\varphi(\gamma|c)d\gamma = P(|Z| < V(c)|c].$$

Actually, Fisher offers the following alternative way of deriving the fiducial density of

$$Z = \frac{\bar{d} - \delta}{\left(\dfrac{s_1^2}{n_1} + \dfrac{s_2^2}{n_2}\right)^{\frac{1}{2}}}.$$

He notes that for $\rho = \sigma_2^2/\sigma_1^2$ known $f\left(\frac{s_2^2}{s_1^2}|\rho\right)$ does not depend on δ and s_2^2/s_1^2 can be considered a "relevant subset" to condition on. Hence he considers the conditional inversion

$$f(z|s_2^2/s_1^2, \rho) = \varphi(z|s_2^2/s_1^2, \rho).$$

Now

$$f(z, s_2^2/s_1^2 | \rho) = f(z | s_2^2/s_1^2, \rho) f(s_2^2/s_1^2 | \rho),$$

so that

$$\varphi(z, \rho) = \varphi(z | s_2^2/s_1^2, \rho) \varphi(\rho | s_2^2/s_1^2),$$

and

$$\int \varphi(z, \rho) d\rho = \varphi(z).$$

Fisher defines a recognizable subset, "In a hypothetical infinite sequence of trials a subset is called recognizable if a rule can be given for identifying each member of that subset before the outcome of the trial. A subset is called relevant if the frequency distribution of the quantity in question depends on the recognizable subset." In view of this Fisher claimed that the Welch solution was defective because it does not recognize the relevant subset s_2^2/s_1^2.

To further complicate matters Buehler and Fedderson (1963) show that a relevant subset exists for the t statistic,

$$t_{n-1} = \frac{(\bar{x} - \mu)\sqrt{n}}{s},$$

which was used and approved by Fisher as not having a recognizable subset. Their argument goes as follows: Consider the case $n = 2$. Here

$$f(t) = \frac{1}{\pi(1 + t^2)},$$

the Cauchy density, so that

$$1 - F(t) = \int_t^\infty f(u) du = \frac{1}{2} - \frac{1}{\pi} \tan^{-1} t.$$

Note that for $t = 1$, $1 - F(t) = \frac{1}{4}$, therefore,

$$0.5 = P[-1 \le t \le 1] = P\left[-1 \le \frac{\frac{x_1 + x_2}{2} - \mu}{s/\sqrt{2}} \le 1 \right]$$

$$= P\left[-1 \le \frac{x_1 + x_2 - 2\mu}{|x_1 - x_2|} \le 1 \right]$$

since

$$s^2 = \frac{1}{2}(x_1 - x_2)^2 \quad \text{or} \quad s = \frac{|x_1 - x_2|}{\sqrt{2}}.$$

Then

$$0.5 = P\left[\frac{x_1 + x_2 - |x_1 - x_2|}{2} \leq \mu \leq \frac{x_1 + x_2 + |x_1 - x_2|}{2}\right] = P[x_{(1)} \leq \mu < x_{(2)}],$$

where $x_{(1)} = \min(x_1, x_2)$ and $x_{(2)} = \max(x_1, x_2)$. Now let

$$A = \{x_1, x_2 : x_{(1)} \leq \mu \leq x_{(2)}\},$$

$$B = \left\{x_1, x_2 : |x_1 - x_2| \geq \frac{2}{3}|x_1 + x_2|\right\}.$$

Then it is shown that B is a relevant subset, that is,

$$P(A|B) \geq 0.5181 \quad \text{and} \quad 1 - P(A|\bar{B}) \leq 0.4819.$$

So whenever B is true the coverage is greater than 0.5 for all μ and less than 0.5 otherwise.

Hence Fisher's fiducial t can be criticized in the same way that Fisher criticized the Welch solution. Fisher had claimed that $\bar{x}$ and s^2 were jointly sufficient for μ and σ^2 and since knowledge of μ and σ^2 is apriori absent, there is no possibility of recognizing any subset of cases within the general set for which any different value the probability should hold.

Statisticians have not been able to adequately resolve these anomalies unto this day.

8.11   THE FIELLER-CREASY PROBLEM

Suppose that we have bivariate random variables (X, Y) with $E(X, Y) = (\mu_x, \mu_y)$ and

$$cov(X, Y) = \begin{pmatrix} \sigma_x^2 & \sigma_{xy} \\ \sigma_{xy} & \sigma_y^2 \end{pmatrix}.$$

We seek a confidence interval for $\mu_y/\mu_x = \alpha$. Further, we assume that $Z = Y - \alpha X$ is $N(0, \sigma^2)$ since $E(Z) = E(Y - \alpha X) = \mu_y - (\mu_y/\mu_x)\mu_x$ and

$$\text{var}(Y - \alpha X) = \sigma_y^2 - 2\alpha\sigma_{xy} + \alpha^2\sigma_x^2 = \sigma^2.$$

Now assume $(X_1, Y_1), \ldots, (X_n, Y_n)$ are i.i.d., then $Z_1, \ldots, Z_n$ is i.i.d. $N(0, \sigma^2)$.

Fieller (1954) presented his solution at a Royal Statistical Society symposium. He observed that

$$\frac{\sqrt{n}\,Z}{s} = \frac{\sqrt{n}(\bar{Y} - \alpha\bar{X})}{\left(s_y^2 - 2\alpha s_{xy} + \alpha^2 s_x^2\right)^{\frac{1}{2}}} = t_{n-1},$$

a student t with $n - 1$ degrees of freedom. Then

$$P\left[\frac{n(\bar{y} - \alpha\bar{x})^2}{s_y^2 - 2\alpha s_{xy} + \alpha^2 s_x^2} \le t_p^2\right] = 1 - 2p, \quad p = \int_{t_p}^{\infty} f(t)dt.$$

Let $\frac{s_y^2}{n} = s_{\bar{y}}^2$, $\frac{s_x^2}{n} = s_{\bar{x}}^2$, $\frac{s_{xy}}{n} = s_{\bar{x}\bar{y}}$, then

$$P\left[\alpha^2\left(\bar{x}^2 - t_p^2 s_{\bar{x}}^2\right) - 2\alpha\left(\bar{x}\bar{y} - s_{\bar{x}\bar{y}} t_p^2\right) + \bar{y}^2 - t_p^2 s_{\bar{y}}^2 \le 0\right] = 1 - 2p,$$

or

$$P\left[\left(\bar{x}^2 - t_p^2 s_{\bar{x}}^2\right)\left(\alpha^2 - \frac{2\alpha\left(\bar{x}\bar{y} - s_{\bar{x}\bar{y}} t_p^2\right)}{\bar{x}^2 - t_p^2 s_{\bar{x}}^2} + \frac{\bar{y}^2 - t_p^2 s_{\bar{y}}^2}{\bar{x}^2 - t_p^2 s_{\bar{x}}^2}\right) \le 0\right]$$

$$= P\left[\left(\bar{x}^2 - t_p^2 s_{\bar{x}}^2\right)(\alpha - \alpha_1)(\alpha - \alpha_2) \le 0\right] = 1 - 2p,$$

where

$$\alpha_1, \alpha_2 = \frac{\dfrac{2\left(\bar{x}\bar{y} - s_{\bar{x}\bar{y}} t_p^2\right)}{\bar{x}^2 - t_p^2 s_{\bar{x}}^2} \pm \sqrt{\dfrac{4\left(\bar{x}\bar{y} - s_{\bar{x}\bar{y}} t_p^2\right)^2}{\left(\bar{x}^2 - t_p^2 s_{\bar{x}}^2\right)^2} - 4\dfrac{\bar{y}^2 - t_p^2 s_{\bar{y}}^2}{\bar{x}^2 - t_p^2 s_{\bar{x}}^2}}}{2}$$

$$= \frac{\left(\bar{x}\bar{y} - s_{\bar{x}\bar{y}} t_p^2\right) \pm \sqrt{\left(\bar{x}\bar{y} - s_{\bar{x}\bar{y}} t_p^2\right)^2 - \left(\bar{y}^2 - t_p^2 s_{\bar{y}}^2\right)\left(\bar{x}^2 - t_p^2 s_{\bar{x}}^2\right)}}{\bar{x}^2 - t_p^2 s_{\bar{x}}^2}.$$

Then

$$P\left[\left(\bar{x}^2 - t_p^2 s_{\bar{x}}^2\right)(\alpha - \alpha_1)(\alpha - \alpha_2) \le 0\right] = 1 - 2p.$$

If the discriminant

$$D = \left(\bar{x}\bar{y} - s_{\bar{x}\bar{y}}t_p^2\right)^2 - \left(\bar{y}^2 - t_p^2 s_{\bar{y}}^2\right)\left(\bar{x}^2 - t_p^2 s_{\bar{x}}^2\right)$$

$$\text{is} \begin{cases} > 0 & \alpha_1 \text{ and } \alpha_2 \text{ real} \\ = 0 & \text{one root } \alpha_1 = \alpha_2 \\ < 0 & \alpha_1 \text{ and } \alpha_2 \text{ are complex roots} \end{cases}$$

Algebra yields

$$D = \left(s_{\bar{x}}^2 s_{\bar{y}}^2 - s_{\bar{x}\bar{y}}^2\right)t_p^2\left(\frac{\bar{x}^2}{s_{\bar{x}}^2} + \frac{\left(\bar{y}s_{\bar{x}}^2 - \bar{x}s_{\bar{x}\bar{y}}\right)^2}{s_{\bar{x}}^2\left(s_{\bar{x}}^2 s_{\bar{y}}^2 - s_{\bar{x}\bar{y}}^2\right)} - t_p^2\right),$$

so for

$$t_p^2 = 0, \quad \alpha_1 = \alpha_2 = \frac{\bar{y}}{\bar{x}}.$$

If

$$t_p^2 = \frac{\bar{x}^2}{s_{\bar{x}}^2} + \frac{\left(\bar{y}s_{\bar{x}}^2 - \bar{x}s_{\bar{x}\bar{y}}\right)^2}{s_{\bar{x}}^2\left(s_{\bar{x}}^2 s_{\bar{y}}^2 - s_{\bar{x}\bar{y}}^2\right)} \equiv \tilde{t}_p^2,$$

$D = 0$ and there is also only one real root $\alpha_1 = \alpha_2 = (\bar{y}s_{\bar{x}\bar{y}} - \bar{x}s_{\bar{y}}^2)/(\bar{y}s_{\bar{y}}^2 - \bar{x}s_{\bar{x}\bar{y}})$.

If $0 < t_p^2 < \tilde{t}_p^2$, there are two real roots since $D > 0$. Otherwise for $D < 0$, there are no real roots. Now for

$$P\left[\left(\bar{x}^2 - t_p^2 s_{\bar{x}}^2\right)(\alpha - \alpha_1)(\alpha - \alpha_2) \leq 0\right] = 1 - 2p,$$

with $\alpha_1 > \alpha_2$, for $\bar{x}^2 - t_p^2 s_{\bar{x}}^2 > 0$, or $t_p^2 < \frac{\bar{x}^2}{s_{\bar{x}}^2}$ then

$$P[\alpha_2 < \alpha < \alpha_1] = 1 - 2p$$

yields inclusive limits.

Now if $\bar{x}^2 - t_p^2 s_{\bar{x}}^2 < 0$, or $t_p^2 > \frac{\bar{x}^2}{s_{\bar{x}}^2}$, and $\alpha_2 > \alpha_1$

$$P(-(\alpha - \alpha_1)(\alpha - \alpha_2) \leq 0] = 1 - 2p,$$

or

$$P[(\alpha_1 - \alpha)(\alpha - \alpha_2) \leq 0] = 1 - 2p = P[\alpha < \alpha_1 \quad \text{or} \quad \alpha > \alpha_2] = 1 - 2p.$$

This yields exclusive limits.

For $\bar{x}^2 = t_p^2 s_{\bar{x}}^2$ we obtain

$$P\left[-2\alpha(\bar{x}\bar{y} - s_{\bar{x}\bar{y}}t_p^2) + \bar{y}^2 - t_p^2 s_{\bar{y}}^2 \leq 0\right] = 1 - 2p$$

$$= P\left[\alpha(s_{\bar{x}\bar{y}}\bar{x}^2 - s_{\bar{x}}^2\bar{x}\bar{y}) \leq (\bar{x}^2 s_{\bar{y}}^2 - \bar{y}^2 s_{\bar{x}}^2)/2\right],$$

by substituting $t_p^2 = \bar{x}^2/s_{\bar{x}}^2$. If $s_{\bar{x}\bar{y}}\bar{x}^2 > s_{\bar{x}}^2\bar{x}\bar{y}$, then

$$1 - 2p = P\left[\alpha \leq \frac{\bar{x}^2 s_{\bar{y}}^2 - \bar{y}^2 s_{\bar{x}}^2}{2(s_{\bar{x}\bar{y}}\bar{x}^2 - s_{\bar{x}}^2\bar{x}\bar{y})}\right] = P\left[\alpha \leq \frac{\bar{y}}{\bar{x}} + \frac{\bar{x}^2 s_{\bar{y}}^2 + \bar{y}^2 s_{\bar{x}}^2 - 2\bar{y}\bar{x}s_{\bar{x}\bar{y}}}{2(s_{\bar{x}\bar{y}}\bar{x}^2 - \bar{x}\bar{y}s_{\bar{x}}^2)}\right]$$

$$= P\left[\alpha \leq \frac{\bar{y}}{\bar{x}} + \frac{(\bar{x}s_{\bar{y}} - \bar{y}s_{\bar{x}})^2 + 2\bar{x}\bar{y}(s_{\bar{y}}s_{\bar{x}} - s_{\bar{x}\bar{y}})}{2(s_{\bar{x}\bar{y}}\bar{x}^2 - s_{\bar{x}}^2\bar{x}\bar{y})}\right],$$

if $\bar{x}\bar{y} > 0$ or

$$P\left[\alpha \leq \frac{\bar{y}}{\bar{x}} + \frac{(\bar{x}s_{\bar{y}} + \bar{y}s_{\bar{x}})^2 - 2\bar{x}\bar{y}(s_{\bar{y}}s_{\bar{x}} - s_{\bar{x}\bar{y}})}{2(s_{\bar{x}\bar{y}}\bar{x}^2 - s_{\bar{x}}^2\bar{x}\bar{y})}\right],$$

if $\bar{x}\bar{y} < 0$. In either case

$$P\left(\alpha \leq \frac{\bar{y}}{\bar{x}} + \Delta\right), \quad \Delta \geq 0.$$

Now if $s_{\bar{x}\bar{y}} < s_{\bar{x}}^2\bar{x}\bar{y}$, then

$$P\left[\alpha \geq \frac{\bar{y}}{\bar{x}} + \frac{(\bar{x}s_{\bar{y}} - \bar{y}s_{\bar{x}})^2 + 2\bar{x}\bar{y}(s_{\bar{y}}s_{\bar{x}} - s_{\bar{x}\bar{y}})}{2(s_{\bar{x}\bar{y}}\bar{x}^2 - s_{\bar{x}}^2\bar{x}\bar{y})}\right],$$

if $\bar{x}\bar{y} > 0$ or

$$P\left[\alpha \geq \frac{\bar{y}}{\bar{x}} + \frac{(\bar{x}s_{\bar{y}} + \bar{y}s_{\bar{x}})^2 - 2\bar{x}\bar{y}(s_{\bar{y}}s_{\bar{x}} - s_{\bar{x}\bar{y}})}{2(s_{\bar{x}\bar{y}}\bar{x}^2 - s_{\bar{x}}^2\bar{x}\bar{y})}\right],$$

if $\bar{x}\bar{y} < 0$ so that

$$P\left[\alpha \geq \frac{\bar{y}}{\bar{x}} - \Delta\right], \quad \Delta \geq 0.$$

Hence $\bar{y}/\bar{x}$ is always included in the interval and one limit goes to infinity.

Fieller claimed his method was a fiducial inversion and Fisher, in commenting at the symposium, agreed. However, it was also clearly a confidence procedure as Neyman asserted at the symposium. Further we have the anomaly that whenever $D < 0$ we have a confidence (fiducial) interval which is the whole real line with coefficient less than 1. Many explanations were proffered for this anomaly, none of which were satisfactory.

Now consider the following simpler version of the problem:
(X, Y) are pairs of independent random variables with $X \sim N(\mu_x, 1)$ and $Y \sim N(\mu_y, 1)$ and let $(X_j, Y_j) j = 1, \ldots, n$ be a random sample with $\alpha = \frac{\mu_y}{\mu_x}$.

Method 1. Confidence interval: Let

$$\alpha\bar{X} - \bar{Y} \sim N\left(0, \frac{1+\alpha^2}{n}\right),$$

since

$$E(\alpha\bar{X} - \bar{Y}) = \alpha\mu_x - \mu_y = 0, \quad \text{var}(\alpha\bar{X} - \bar{Y}) = \frac{\alpha^2}{n} + \frac{1}{n} = \frac{\alpha^2+1}{n}.$$

Then

$$P\left(\frac{(\alpha\bar{x} - \bar{y})^2 n}{1 + \alpha^2} \le u_{2p}\right) = 1 - 2p,$$

where

$$2p = \int_{u_{2p}}^{\infty} f_U(u)du,$$

where U is χ^2 with one degree of freedom and

$$1 - 2p = P\left[\alpha^2\bar{x}^2 - 2\alpha\bar{x}\bar{y} + \bar{y}^2 - \frac{(1 + \alpha^2)}{n}u_{2p} \le 0\right].$$

Again, this can be rewritten as

$$1 - 2p = P\left[\alpha^2\left(\bar{x}^2 - \frac{u_{2p}}{n}\right) - 2\alpha\bar{x}\bar{y} + \left(\bar{y}^2 - \frac{u_{2p}}{n}\right) \le 0\right]$$

$$= P\left[\left(\bar{x}^2 - \frac{u_{2p}}{n}\right)(\alpha - \alpha_1)(\alpha - \alpha_2) \le 0\right],$$

where

$$\alpha_1, \alpha_2 = \left[\bar{x}\bar{y} \pm \sqrt{\bar{x}^2\bar{y}^2 - \left(\bar{x}^2 - \frac{u_{2p}}{n}\right)\left(\bar{y}^2 - \frac{u_{2p}}{n}\right)}\right] \Big/ \left(\bar{x}^2 - \frac{u_{2p}}{n}\right)$$

$$= \left[\bar{x}\bar{y} \pm \sqrt{\frac{u_{2p}}{n}\left(\bar{x}^2 + \bar{y}^2 - \frac{u_{2p}}{n}\right)}\right] \Big/ \left(\bar{x}^2 - \frac{u_{2p}}{n}\right).$$

Then

$$D = 0 \quad \text{if } u_{2p} = 0 \quad \text{or} \quad u_{2p} = n(\bar{x}^2 + \bar{y}^2) \text{ one root, as before}$$

$$D > 0, \quad 0 < u_{2p} < n(\bar{x}^2 + \bar{y}^2) \quad \alpha_1 \text{ and } \alpha_2 \text{ real}$$

$$D < 0, \quad u_{2p} > n(\bar{x}^2 + \bar{y}^2) \quad \alpha_1 \text{ and } \alpha_2 \text{ complex}$$

So we have the same situation as before with confidence limits (also presumably fiducial limits) being inclusive or exclusive or when $u_{2p} > n(\bar{x}^2 + \bar{y}^2)$ the whole real line with confidence coefficient $1 - 2p$.

However, Creasy (1954), at the same symposium, offered what appears to be another "fiducial" solution along the following lines:

Method 2. Fiducial interval: Consider the simple example as before, where

$$f(\bar{x}|\mu_x, n^{-1}) = \sqrt{\frac{n}{2\pi}} e^{-\frac{n}{2}(\bar{x}-\mu_x)^2}, \quad \text{independent of}$$

$$f(\bar{y}|\mu_y, n^{-1}) = \sqrt{\frac{n}{2\pi}} e^{\frac{n}{2}(\bar{y}-\mu_y)^2},$$

from which we easily get the joint fiducial density,

$$\varphi(\mu_x, \mu_y) = \frac{n}{2\pi} e^{-\frac{n}{2}(\mu_x-\bar{x})^2 - \frac{n}{2}(\mu_y-\bar{y})^2}.$$

Let $\alpha = \mu_y/\mu_x$ or $\alpha\mu_x = \mu_y$, then

$$\left| \frac{\partial \mu_y}{\partial \alpha} \right| = |\mu_x|,$$

so that

$$\varphi(\mu_x, \alpha) = |\mu_x| \frac{n}{2\pi} e^{-\frac{n}{2}(\mu_x-\bar{x})^2 - \frac{n}{2}(\alpha\mu_x-\bar{y})^2}.$$

Integrate out μ_x and obtain

$$\varphi(\alpha) = \int_{-\infty}^{\infty} \varphi(\mu_x, \alpha) d\mu_x = \int_0^{\infty} \frac{n\mu_x}{2\pi} e^{-\frac{n}{2}[\mu_x^2 - 2\bar{x}\mu_x + \bar{x}^2 + \alpha^2\mu_x^2 - 2\alpha\mu_x\bar{y} + \bar{y}^2]} d\mu_x$$

$$- \int_{-\infty}^{0} \frac{n\mu_x}{2\pi} e^{-\frac{n}{2}[\mu_x^2 - 2\bar{x}\mu_x + \bar{x}^2 + \alpha^2\mu_x^2 - 2\alpha\mu_x\bar{y} + \bar{y}^2]} d\mu_x.$$

Add and subtract $\displaystyle\int_{-\infty}^{0}\varphi(\mu_x, \alpha)d\mu_x$ so that

$$\varphi(\alpha) = \int_{-\infty}^{\infty}\frac{n\mu_x}{2\pi}e^{-\frac{n}{2}[\mu_x^2 - 2\bar{x}\mu_x + \bar{x}^2 + \alpha^2\mu_x^2 - 2\alpha\mu_x\bar{y} + \bar{y}^2]}d\mu_x$$

$$- 2\int_{-\infty}^{0}\frac{n\mu_x}{2\pi}e^{-\frac{n}{2}[\mu_x^2 - 2\bar{x}\mu_x + \bar{x}^2 + \alpha^2\mu_x^2 - 2\alpha\mu_x\bar{y} + \bar{y}^2]}d\mu_x.$$

Now

$$e^{-\frac{n}{2}(\mu_x^2(1+\alpha^2) - 2\mu_x(\bar{x}+\alpha\bar{y}) + \bar{x}^2 + \bar{y}^2)} = e^{-n\frac{(\bar{x}^2+\bar{y}^2)}{2}}e^{-\frac{n(1+\alpha^2)}{2}(\mu_x^2 - 2\mu_x\frac{(\bar{x}+\alpha\bar{y})}{1+\alpha^2})}$$

$$= e^{\frac{n}{2}\frac{(\bar{x}+\alpha\bar{y})^2}{1+\alpha^2}}e^{-\frac{n}{2}(\bar{x}^2+\bar{y}^2)}e^{-\frac{n(1+\alpha^2)}{2}(\mu_x - \frac{\bar{y}\alpha+\bar{x}}{1+\alpha^2})^2},$$

so

$$\varphi(\alpha) = \frac{ne^{\frac{n}{2}\frac{(\bar{x}+\alpha\bar{y})^2}{1+\alpha^2} - \frac{n}{2}(\bar{x}^2+\bar{y}^2)}}{2\pi}\left[\int_{-\infty}^{\infty}\mu_x e^{-\frac{n(1+\alpha^2)}{2}(\mu_x - \frac{\bar{y}\alpha+\bar{x}}{1+\alpha^2})^2}d\mu_x\right.$$

$$\left. - 2\int_{-\infty}^{0}\left[\mu_x e^{-\frac{n(1+\alpha^2)}{2}(\mu_x - \frac{\bar{y}\alpha+\bar{x}}{1+\alpha^2})^2}\right]d\mu_x\right].$$

Let

$$t = \left(\mu_x - \frac{\bar{y}\alpha + \bar{x}}{1 + \alpha^2}\right)\sqrt{(1 + \alpha^2)n} = \mu_x\sqrt{(1+\alpha^2)n} - \frac{(\bar{y}\alpha + \bar{x})\sqrt{n}}{\sqrt{1+\alpha^2}},$$

$$r = \frac{(\bar{y}\alpha + \bar{x})\sqrt{n}}{\sqrt{1+\alpha^2}} \quad\text{or}\quad \mu_x = \frac{(t+r)}{\sqrt{(1+\alpha^2)n}}.$$

Then

$$\varphi(\alpha) = \frac{e^{-\frac{n(\bar{y}-\alpha\bar{x})^2}{2(1+\alpha^2)}}}{\sqrt{2\pi}(1+\alpha^2)}\frac{1}{\sqrt{2\pi}}\left[\int_{-\infty}^{\infty}(t+r)e^{-\frac{t^2}{2}}dt - 2\int_{-\infty}^{-r}(t+r)e^{-\frac{t^2}{2}}dt\right].$$

Now consider

$$\frac{1}{\sqrt{2\pi}}\int_{-\infty}^{\infty}(t+r)e^{-\frac{t^2}{2}}dt - \frac{2}{\sqrt{2\pi}}\int_{-\infty}^{-r}(t+r)e^{-\frac{t^2}{2}}dt$$

$$= r - \frac{2}{\sqrt{2\pi}}\int_{-\infty}^{-r}te^{-\frac{t^2}{2}}dt - 2r\int_{-\infty}^{-r}\frac{e^{-t^2/2}}{\sqrt{2\pi}}dt$$

$$= r - 2r\Phi(-r) - \frac{2}{\sqrt{2\pi}}\int_{-\infty}^{-r}te^{-\frac{t^2}{2}}dt$$

$$= r(1 - 2\Phi(-r)) + \frac{2}{\sqrt{2\pi}}e^{-r^2/2},$$

so that

$$\varphi(\alpha) = \frac{e^{-\frac{n(\bar{y} - \alpha\bar{x})^2}{2(1+\alpha^2)}}}{(1+\alpha^2)\sqrt{2\pi}} \left[\frac{2e^{-r^2/2}}{\sqrt{2\pi}} + \frac{r}{\sqrt{2\pi}} \int_{-r}^{r} e^{-t^2/2} dt \right].$$

Let

$$\tan\theta = \frac{\bar{x}\alpha - \bar{y}}{\bar{x} + \alpha\bar{y}}, \qquad \theta = \tan^{-1}\frac{\bar{x}\alpha - \bar{y}}{\bar{x} + \alpha\bar{y}}, \qquad -\frac{\pi}{2} \le \theta \le \frac{\pi}{2}, \qquad a = \sqrt{(\bar{x}^2 + \bar{y}^2)n}.$$

Then

$$\cos\theta = \frac{\bar{x} + \alpha\bar{y}}{\sqrt{(\bar{x}^2 + \bar{y}^2)(1 + \alpha^2)}}, \qquad \sin\theta = \frac{\bar{x}\alpha - \bar{y}}{\sqrt{(\bar{x}^2 + \bar{y}^2)(1 + \alpha^2)}},$$

and

$$a\cos\theta = \frac{(\bar{x} + \alpha\bar{y})\sqrt{n}}{\sqrt{1 + \alpha^2}} = r, \qquad a\sin\theta = \frac{(\bar{x}\alpha - \bar{y})\sqrt{n}}{\sqrt{1 + \alpha^2}}.$$

Recall the trigonometric identity

$$\tan(c \pm b) = \frac{\tan c \pm \tan b}{1 \mp \tan c \tan b}, \qquad \text{so} \qquad \theta = \tan^{-1}\frac{\bar{x}\alpha - \bar{y}}{\bar{x} + \alpha\bar{y}} = \tan^{-1}\alpha - \tan^{-1}\frac{\bar{y}}{\bar{x}}.$$

Therefore,

$$d\theta = \frac{d\alpha}{1 + \alpha^2}$$

and

$$\varphi(\theta) = \frac{e^{-\frac{a^2}{2}\sin^2\theta}}{\sqrt{2\pi}} \left[\frac{2e^{-\frac{a^2\cos^2\theta}{2}}}{\sqrt{2\pi}} + \frac{a\cos\theta}{\sqrt{2\pi}} \int_{-a\cos\theta}^{a\cos\theta} e^{-t^2/2} dt \right]$$

$$= \frac{1}{\pi} e^{-a^2/2} + \frac{a\cos\theta}{\pi} e^{-\frac{a^2\sin^2\theta}{2}} \int_{0}^{a\cos\theta} e^{-t^2/2} dt.$$

Hence,

$$\int_{0}^{a\cos\theta} e^{-t^2/2} dt = e^{-\frac{a^2\cos^2\theta}{2}} \left[\frac{a\cos\theta}{1} + \frac{(a\cos\theta)^3}{1\cdot 3} + \frac{(a\cos\theta)^5}{1\cdot 3\cdot 5} + \cdots \right],$$

so that

$$\varphi(\theta) = \frac{e^{-a^2/2}}{\pi} \left[1 + \frac{(a\cos\theta)^2}{1} + \frac{(a\cos\theta)^4}{1\cdot3} + \frac{(a\cos\theta)^6}{1\cdot3\cdot5} + \cdots \right].$$

Note that $\varphi(\theta) = \varphi(-\theta)$ since $\cos\theta = \cos(-\theta)$. Let

$$\int_{\theta_0}^{\pi/2} \varphi(\theta)d\theta = p.$$

Then

$$P[-\theta_0 \le \theta \le \theta_0] = 1 - 2p$$

or

$$P\left[-\theta_0 \le \tan^{-1}\frac{\bar{x}\alpha - \bar{y}}{\bar{x} + \alpha\bar{y}} \le \theta_0 \right] = 1 - 2p$$

$$= P\left[-\theta_0 \le \tan^{-1}\alpha - \tan^{-1}\frac{\bar{y}}{\bar{x}} \le \theta_0 \right]$$

$$= P\left[-\theta_0 + \tan^{-1}\frac{\bar{y}}{\bar{x}} \le \tan^{-1}\alpha \le \theta_0 + \tan^{-1}\frac{\bar{y}}{\bar{x}} \right].$$

Now

$$\alpha_1 = \tan\left[\tan^{-1}\frac{\bar{y}}{\bar{x}} - \theta_0 \right] = \left(\frac{\bar{y}}{\bar{x}} - \tan\theta_0 \right) \Big/ \left(1 + \frac{\bar{y}}{\bar{x}}\tan\theta_0 \right)$$

$$\alpha_2 = \tan\left[\tan^{-1}\frac{\bar{y}}{\bar{x}} + \theta_0 \right] = \left(\frac{\bar{y}}{\bar{x}} + \tan\theta_0 \right) \Big/ \left(1 - \frac{\bar{y}}{\bar{x}}\tan\theta_0 \right),$$

because

$$\tan(c \pm b) = \frac{\tan c \pm \tan b}{1 \mp \tan c\tan b}.$$

If $\alpha_2 > \alpha_1$ then we have inclusive limits

$$P[\alpha_1 \le \alpha \le \alpha_2] = 1 - 2p.$$

If $\alpha_1 > \alpha_2$ then $P[\alpha \ge \alpha_1 \text{ or } \alpha \le \alpha_2] = 1 - 2p$ and limits are exclusive, but for $0 < p < \frac{1}{2}$ we never have the whole real line.

Calculations made by Creasey for the comparison of the two latter cases show that for the same $1 - 2p$, Creasey's limits are included in Fieller's limits.

REFERENCES

Basu, D. (1964). Recovery of ancillary information. *Sankhya, A*, **26**, 3–26.

Buehler, R. J. and Feddersen, A. P. (1963). Note on a conditional property of Student's *t*. *Annals of Mathematical Statistics*, **34**, 1098–1100.

Cox, D. R. (1971). The choice between alternative ancillary statistics, *Journal of the Royal Statistical Society, B*, **2**, 251–255.

Creasy, M. A. (1954). Limits for the ratio of means. *Journal of the Royal Statistical Society, B*, **16(2)**, 186–194.

Fieller, E. C. (1954). Some problems in interval estimation. *Journal of the Royal Statistical Society, B*, **16(2)**, 175–185.

Fisher, R. A. (1956a). On a test of significance in Pearson's Biometrika Tables (No. 11). *Journal of the Royal Statistical Society, B*, **18(1)**, 56–60.

Fisher, R. A. (1956b). *Statistical Methods and Scientific Inference*. Oliver and Boyd, Edinburgh.

Lindley, D. V. (1958). Fiducial inference and Bayes' theorem. *Journal of the Royal Statistical Society, B*, **20**, 102–107.

Pearson, E. and Hartley, H. O. (1966). *Biometrika Tables for Statisticians*. **Vol. 1**, 142.

Scheffé, H. (1943). On solutions of the Behrens-Fisher problem based on the *t*-distribution. *Annals of Mathematical Statistics*, **14**, 35–44.

Welch, B. L. (1947). The generalization of 'Students'' problems when several different population variances are involved. *Biometrika*, **34**, 28–35.

Welch, B. L. (1956). Note on some criticisms made by Sir Ronald Fisher. *Journal of the Royal Statistical Society, B*, **18(1)**, 297–302.

References

Arbuthnot, J. (1710). Argument for divine providence taken from the constant regularity of the births of both sexes. *Philosophical Transactions of the Royal Society*, **XXIII**, 186–190.

Barnard, G. A. (1945). A new test for 2×2 tables. *Nature*, **156**, 177.

Barnard, G. A. (1969). Practical application of tests of power one. *Bulletin of the International Statistical Institute*, **XLIII**, 1, 389–393.

Barnard, G. A. (1974). Can we all agree what we mean by estimation? *Utilitas Mathematica*, **6**, 3–22.

Barnard, G. A., Jenkins, G. M. and Winsten, C. B. (1962). Likelihood inference and time series. *Journal of the Royal Statistical Society, A*, **125**, 321–372.

Barnard, G. A. and Sprott, D. A. (1971). A note on Basu's examples of anomalous ancillary statistics (see reply, p 175). *Foundations of Statistical Inference*, Godambe, V. P. and Sprott, D. A., eds. New York: Holt, Rinehart and Winston, pp. 163–170.

Basu, D. (1964). Recovery of ancillary information. *Sankhya, A*, **26**, 3–26.

Bernardo, J. M. and Smith, A. F. M. (1994). *Bayesian Theory*, New York: Wiley.

Bhattacharyya, A. (1946). On some analogues of the amount of information and their use in statistical estimation. *Sankhya*, **8**, 1–14, 201–218, 315–328.

Birnbaum, A. (1962). On the foundations of statistical inference. *Journal of the American Statistical Association*, **57**, 269–326.

Buehler, R. J. and Feddersen, A. P. (1963). Note on a conditional property of Student's *t*. *Annals of Mathematical Statistics*, **34**, 1098–1100.

Burbea, J. and Rao, C. R. (1982). Entropy differential metric, distance and divergent measures in probability spaces: a unified approach. *Journal of Multivariate Analysis*, **12**, 575–596.

Cornfield, J. (1969). The Bayesian outlook and its applications. *Biometrics*, **25(4)**, 617–657.

Cox, D. R. (1971). The choice between alternative ancillary statistics, *Journal of the Royal Statistical Society, B*, **2**, 251–255.

Cox, D. R. and Hinkley, D. V. (1974). *Theoretical Statistics*. London: Chapman and Hall.

Creasy, M. A. (1954). Limits for the ratio of means. *Journal of the Royal Statistical Society, B*, **16(2)**, 186–194.

de Finetti, B. (1973). Le Prevision: ses lois logiques, ses sources subjectives. *Annals Institute Poincaire, tome VIII*, fasc. 1, 1–68. Reprinted in Studies in Subjective Probability. Melbourne, FL: Krieger, 1980 (English translation).

Dublin, T. et al. (1964). Red blood cell groups and ABH secretor system as genetic indicators of susceptibility to rheumatic fever and rheumatic heart disease. *British Medical Journal*, September 26, **ii**, 775–779.

Edwards, A. W. F. (1992). *Likelihood*. Baltimore: John Hopkins University Press.

Feller, W. (1966). *Introduction to Probability Theory and Its Applications*. Volume II. New York: Wiley.

Fieller, E. C. (1954). Some problems in interval estimation. *Journal of the Royal Statistical Society, B*, **16(2)**, 175–185.

Fisher, R. A. (1935). *The Design of Experiments*. Edinburgh: Oliver and Boyd.

Fisher, R. A. (1956a). On a test of significance in Pearson's *Biometrika Tables* (No. 11). *Journal of the Royal Statistical Society, B*, **18(1)**, 56–60.

Fisher, R. A. (1956b). *Statistical Methods and Scientific Inference*. Edinburgh: Oliver and Boyd.

Freedman, D. A. and Purves, R. A. (1969). Bayes' method for bookies. *Annals of Mathematical Statistics*, **40**, 1177–1186.

Geisser, S. (1969). Discussion of Bayesian outlook and its application. *Biometrics*, 643–645.

Hacking, I. (1965). *Logic of Statistical Inference*. Cambridge: Cambridge University Press.

Heath, D. and Sudderth, W. D. (1976). de Finetti's theorem on exchangeable variables. *American Statistician*, **7(4)**, 718–728.

Jeffreys, H. (1961). *Theory of Probability*. Oxford: Clarendon Press.

Kalbfleisch, J. D. and Sprott, D. A. (1970). Application of likelihood methods to models involving large numbers of parameters. *Journal of the Royal Statistical Society, B*, **32**, 175–208.

Lehmann, E. L. (1950). Some principles of the theory of testing hypothesis. *Annals of Mathematical Statistics*, **21**, 1–26.

Lehmann, E. L. (1959). *Testing Statistical Hypothesis*, New York: Wiley.

Lehmann, E. L. (1983). *Theory of Point Estimation*. New York: Wiley.

Lindley, D. V. (1958). Fiducial inference and Bayes' theorem. *Journal of the Royal Statistical Society, B*, **20**, 102–107.

Loeve, M. (1960). *Probability Theory*. New York: Van Nostrand.

Neyman, J. and Pearson, E. S. (1933). On the problem of the most efficient test of statistical hypothesis. *Philosophical Transactions of the Royal Society*, **LCXXXI**, 289–337.

Neyman, J. and Pearson, E. S. (1936–1938). Contributions to the theory of testing statistical hypotheses. *Statistical Research Memoirs*. **I**, 1–37; **II**, 25–57.

Pearson, E. and Hartley, H. O. (1966). *Biometrika Tables for Statisticians*. **1**, 142.

Pitman, E. J. G. (1979). *Some Basic Theory for Statistical Inference*. New York: Chapman and Hall.

Rabinovitch, N. L. (1970). Studies in the history of probability and statistics, XXIV Combinations and probability in rabbinic literature. *Biometrika*, **57**, 203–205.

Royall, R. M. (1997). *Statistical Evidence: A Likelihood Paradigm*. New York: Chapman and Hall.

Scheffé, H. (1943). On solutions of the Behrens-Fisher problem based on the *t*-distribution. *Annals of Mathematical Statistics*, **14**, 35–44.

Stein, C. M. (1951). A property of some tests of composite hypothesis. *Annals of Mathematical Statistics*, **22**, 475–476.

Wald, A. (1947). *Sequential Analysis*. New York: Wiley.

Welch, B. L. (1947). The generalization of 'Students'' problems when several different population variances are involved. *Biometrika*, **34**, 28–35.

Welch, B. L. (1956). Note on some criticisms made by Sir Ronald Fisher. *Journal of the Royal Statistical Society, B*, **18(1)**, 297–302.

Index

Modes of Parametric Statistical Inference, by Seymour Geisser
Copyright © 2006 John Wiley & Sons, Inc.

WILEY SERIES IN PROBABILITY AND STATISTICS
ESTABLISHED BY WALTER A. SHEWHART AND SAMUEL S. WILKS

Editors: David J. Balding, Noel A. C. Cressie, Nicholas I. Fisher,
Iain M. Johnstone, J. B. Kadane,Geert Molenberghs. Louise M. Ryan,
David W. Scott, Adrian F. M. Smith, Jozef L. Teugels
Editors Emeriti: Vic Barnett, J. Stuart Hunter, David G. Kendall

The *Wiley Series in Probability and Statistics* is well established and authoritative. It covers many topics of current research interest in both pure and applied statistics and probability theory. Written by leading statisticians and institutions, the titles span both state-of-the-art developments in the field and classical methods.

Reflecting the wide range of current research in statistics, the series encompasses applied, methodological and theoretical statistics, ranging from applications and new techniques made possible by advances in computerized practice to rigorous treatment of theoretical approaches.

This series provides essential and invaluable reading for all statisticians, whether in academia, industry, government, or research.

*Now available in a lower priced paperback edition in the Wiley Classics Library.

BASU and RIGDON · Statistical Methods for the Reliability of Repairable Systems

BATES and WATTS · Nonlinear Regression Analysis and Its Applications

BECHHOFER, SANTNER, and GOLDSMAN · Design and Analysis of Experiments for Statistical Selection, Screening, and Multiple Comparisons

BELSLEY · Conditioning Diagnostics: Collinearity and Weak Data in Regression

† BELSLEY, KUH, and WELSCH · Regression Diagnostics: Identifying Influential Data and Sources of Collinearity

BENDAT and PIERSOL · Random Data: Analysis and Measurement Procedures, *Third Edition*

BERRY, CHALONER, and GEWEKE · Bayesian Analysis in Statistics and Econometrics: Essays in Honor of Arnold Zellner

BERNARDO and SMITH · Bayesian Theory

BHAT and MILLER · Elements of Applied Stochastic Processes, *Third Edition*

BHATTACHARYA and WAYMIRE · Stochastic Processes with Applications

† BIEMER, GROVES, LYBERG, MATHIOWETZ, and SUDMAN · Measurement Errors in Surveys

BILLINGSLEY · Convergence of Probability Measures, *Second Edition*

BILLINGSLEY · Probability and Measure, *Third Edition*

BIRKES and DODGE · Alternative Methods of Regression

BLISCHKE AND MURTHY (editors) · Case Studies in Reliability and Maintenance

BLISCHKE AND MURTHY · Reliability: Modeling, Prediction, and Optimization

BLOOMFIELD · Fourier Analysis of Time Series: An Introduction, *Second Edition*

BOLLEN · Structural Equations with Latent Variables

BOROVKOV · Ergodicity and Stability of Stochastic Processes

BOULEAU · Numerical Methods for Stochastic Processes

BOX · Bayesian Inference in Statistical Analysis

BOX · R. A. Fisher, the Life of a Scientist

BOX and DRAPER · Empirical Model-Building and Response Surfaces

* BOX and DRAPER · Evolutionary Operation: A Statistical Method for Process Improvement

BOX, HUNTER, and HUNTER · Statistics for Experimenters: Design, Innovation, and Discovery, *Second Editon*

BOX and LUCEÑO · Statistical Control by Monitoring and Feedback Adjustment

BRANDIMARTE · Numerical Methods in Finance: A MATLAB-Based Introduction

BROWN and HOLLANDER · Statistics: A Biomedical Introduction

BRUNNER, DOMHOF, and LANGER · Nonparametric Analysis of Longitudinal Data in Factorial Experiments

BUCKLEW · Large Deviation Techniques in Decision, Simulation, and Estimation

CAIROLI and DALANG · Sequential Stochastic Optimization

CASTILLO, HADI, BALAKRISHNAN, and SARABIA · Extreme Value and Related Models with Applications in Engineering and Science

CHAN · Time Series: Applications to Finance

CHARALAMBIDES · Combinatorial Methods in Discrete Distributions

CHATTERJEE and HADI · Sensitivity Analysis in Linear Regression

CHATTERJEE and PRICE · Regression Analysis by Example, *Third Edition*

CHERNICK · Bootstrap Methods: A Practitioner's Guide

CHERNICK and FRIIS · Introductory Biostatistics for the Health Sciences

CHILÈS and DELFINER · Geostatistics: Modeling Spatial Uncertainty

CHOW and LIU · Design and Analysis of Clinical Trials: Concepts and Methodologies, *Second Edition*

CLARKE and DISNEY · Probability and Random Processes: A First Course with Applications, *Second Edition*

* COCHRAN and COX · Experimental Designs, *Second Edition*

---

*Now available in a lower priced paperback edition in the Wiley Classics Library.

†Now available in a lower priced paperback edition in the Wiley–Interscience Paperback Series.

*Now available in a lower priced paperback edition in the Wiley Classics Library.

*Now available in a lower priced paperback edition in the Wiley Classics Library.

†Now available in a lower priced paperback edition in the Wiley–Interscience Paperback Series.

JOHNSON and BHATTACHARYYA · Statistics: Principles and Methods, *Fifth Edition*

JOHNSON and KOTZ · Distributions in Statistics

JOHNSON and KOTZ (editors) · Leading Personalities in Statistical Sciences: From the Seventeenth Century to the Present

JOHNSON, KOTZ, and BALAKRISHNAN · Continuous Univariate Distributions, Volume 1, *Second Edition*

JOHNSON, KOTZ, and BALAKRISHNAN · Continuous Univariate Distributions, Volume 2, *Second Edition*

JOHNSON, KOTZ, and BALAKRISHNAN · Discrete Multivariate Distributions

JOHNSON, KOTZ, and KEMP · Univariate Discrete Distributions, *Second Edition*

JUDGE, GRIFFITHS, HILL, LÜTKEPOHL, and LEE · The Theory and Practice of Econometrics, *Second Edition*

JUREČ KOVÁ and SEN · Robust Statistical Procedures: Aymptotics and Interrelations

JUREK and MASON · Operator-Limit Distributions in Probability Theory

KADANE · Bayesian Methods and Ethics in a Clinical Trial Design

KADANE AND SCHUM · A Probabilistic Analysis of the Sacco and Vanzetti Evidence

KALBFLEISCH and PRENTICE · The Statistical Analysis of Failure Time Data, *Second Edition*

KASS and VOS · Geometrical Foundations of Asymptotic Inference

† KAUFMAN and ROUSSEEUW · Finding Groups in Data: An Introduction to Cluster Analysis

KEDEM and FOKIANOS · Regression Models for Time Series Analysis

KENDALL, BARDEN, CARNE, and LE · Shape and Shape Theory

KHURI · Advanced Calculus with Applications in Statistics, *Second Edition*

KHURI, MATHEW, and SINHA · Statistical Tests for Mixed Linear Models

* KISH · Statistical Design for Research

KLEIBER and KOTZ · Statistical Size Distributions in Economics and Actuarial Sciences

KLUGMAN, PANJER, and WILLMOT · Loss Models: From Data to Decisions, *Second Edition*

KLUGMAN, PANJER, and WILLMOT · Solutions Manual to Accompany Loss Models: From Data to Decisions, *Second Edition*

KOTZ, BALAKRISHNAN, and JOHNSON · Continuous Multivariate Distributions, Volume 1, *Second Edition*

KOTZ and JOHNSON (editors) · Encyclopedia of Statistical Sciences: Volumes 1 to 9 with Index

KOTZ and JOHNSON (editors) · Encyclopedia of Statistical Sciences: Supplement Volume

KOTZ, READ, and BANKS (editors) · Encyclopedia of Statistical Sciences: Update Volume 1

KOTZ, READ, and BANKS (editors) · Encyclopedia of Statistical Sciences: Update Volume 2

KOVALENKO, KUZNETZOV, and PEGG · Mathematical Theory of Reliability of Time-Dependent Systems with Practical Applications

LACHIN · Biostatistical Methods: The Assessment of Relative Risks

LAD · Operational Subjective Statistical Methods: A Mathematical, Philosophical, and Historical Introduction

LAMPERTI · Probability: A Survey of the Mathematical Theory, *Second Edition*

LANGE, RYAN, BILLARD, BRILLINGER, CONQUEST, and GREENHOUSE · Case Studies in Biometry

LARSON · Introduction to Probability Theory and Statistical Inference, *Third Edition*

LAWLESS · Statistical Models and Methods for Lifetime Data, *Second Edition*

LAWSON · Statistical Methods in Spatial Epidemiology

LE · Applied Categorical Data Analysis

LE · Applied Survival Analysis

LEE and WANG · Statistical Methods for Survival Data Analysis, *Third Edition*

LEPAGE and BILLARD · Exploring the Limits of Bootstrap

LEYLAND and GOLDSTEIN (editors) · Multilevel Modelling of Health Statistics

*Now available in a lower priced paperback edition in the Wiley Classics Library.

†Now available in a lower priced paperback edition in the Wiley–Interscience Paperback Series.

PANKRATZ · Forecasting with Univariate Box-Jenkins Models: Concepts and Cases
* PARZEN · Modern Probability Theory and Its Applications
PEÑA, TIAO, and TSAY · A Course in Time Series Analysis
PIANTADOSI · Clinical Trials: A Methodologic Perspective
PORT · Theoretical Probability for Applications
POURAHMADI · Foundations of Time Series Analysis and Prediction Theory
PRESS · Bayesian Statistics: Principles, Models, and Applications
PRESS · Subjective and Objective Bayesian Statistics, *Second Edition*
PRESS and TANUR · The Subjectivity of Scientists and the Bayesian Approach
PUKELSHEIM · Optimal Experimental Design
PURI, VILAPLANA, and WERTZ · New Perspectives in Theoretical and Applied Statistics
† PUTERMAN · Markov Decision Processes: Discrete Stochastic Dynamic Programming QIU · Image Processing and Jump Regression Analysis
* RAO · Linear Statistical Inference and Its Applications, *Second Edition*
RAUSAND and HØYLAND · System Reliability Theory: Models, Statistical Methods, and Applications, *Second Edition*
RENCHER · Linear Models in Statistics
RENCHER · Methods of Multivariate Analysis, *Second Edition*
RENCHER · Multivariate Statistical Inference with Applications
* RIPLEY · Spatial Statistics
RIPLEY · Stochastic Simulation
ROBINSON · Practical Strategies for Experimenting
ROHATGI and SALEH · An Introduction to Probability and Statistics, *Second Edition*
ROLSKI, SCHMIDLI, SCHMIDT, and TEUGELS · Stochastic Processes for Insurance and Finance
ROSENBERGER and LACHIN · Randomization in Clinical Trials: Theory and Practice
ROSS · Introduction to Probability and Statistics for Engineers and Scientists
† ROUSSEEUW and LEROY · Robust Regression and Outlier Detection
* RUBIN · Multiple Imputation for Nonresponse in Surveys
RUBINSTEIN · Simulation and the Monte Carlo Method
RUBINSTEIN and MELAMED · Modern Simulation and Modeling
RYAN · Modern Regression Methods
RYAN · Statistical Methods for Quality Improvement, *Second Edition*
SALTELLI, CHAN, and SCOTT (editors) · Sensitivity Analysis
* SCHEFFE · The Analysis of Variance
SCHIMEK · Smoothing and Regression: Approaches, Computation, and Application
SCHOTT · Matrix Analysis for Statistics, *Second Edition*
SCHOUTENS · Levy Processes in Finance: Pricing Financial Derivatives
SCHUSS · Theory and Applications of Stochastic Differential Equations
SCOTT · Multivariate Density Estimation: Theory, Practice, and Visualization
* SEARLE · Linear Models
SEARLE · Linear Models for Unbalanced Data
SEARLE · Matrix Algebra Useful for Statistics
SEARLE, CASELLA, and McCULLOCH · Variance Components
SEARLE and WILLETT · Matrix Algebra for Applied Economics
SEBER and LEE · Linear Regression Analysis, *Second Edition*
† SEBER · Multivariate Observations
† SEBER and WILD · Nonlinear Regression
SENNOTT · Stochastic Dynamic Programming and the Control of Queueing Systems
* SERFLING · Approximation Theorems of Mathematical Statistics
SHAFER and VOVK · Probability and Finance: It's Only a Game!

SILVAPULLE and SEN · Constrained Statistical Inference: Inequality, Order, and Shape Restrictions

SMALL and MCLEISH · Hilbert Space Methods in Probability and Statistical Inference

SRIVASTAVA · Methods of Multivariate Statistics

STAPLETON · Linear Statistical Models

STAUDTE and SHEATHER · Robust Estimation and Testing

STOYAN, KENDALL, and MECKE · Stochastic Geometry and Its Applications, *Second Edition*

STOYAN and STOYAN · Fractals, Random Shapes and Point Fields: Methods of Geometrical Statistics

STYAN · The Collected Papers of T. W. Anderson: 1943–1985

SUTTON, ABRAMS, JONES, SHELDON, and SONG · Methods for Meta-Analysis in Medical Research

TANAKA · Time Series Analysis: Nonstationary and Noninvertible Distribution Theory

THOMPSON · Empirical Model Building

THOMPSON · Sampling, *Second Edition*

THOMPSON · Simulation: A Modeler's Approach

THOMPSON and SEBER · Adaptive Sampling

THOMPSON, WILLIAMS, and FINDLAY · Models for Investors in Real World Markets

TIAO, BISGAARD, HILL, PEÑA, and STIGLER (editors) · Box on Quality and
Discovery: with Design, Control, and Robustness

TIERNEY · LISP-STAT: An Object-Oriented Environment for Statistical Computing and Dynamic Graphics

TSAY · Analysis of Financial Time Series

UPTON and FINGLETON · Spatial Data Analysis by Example, Volume II: Categorical and Directional Data

VAN BELLE · Statistical Rules of Thumb

VAN BELLE, FISHER, HEAGERTY, and LUMLEY · Biostatistics: A Methodology for the Health Sciences, *Second Edition*

VESTRUP · The Theory of Measures and Integration

VIDAKOVIC · Statistical Modeling by Wavelets

VINOD and REAGLE · Preparing for the Worst: Incorporating Downside Risk in Stock Market Investments

WALLER and GOTWAY · Applied Spatial Statistics for Public Health Data

WEERAHANDI · Generalized Inference in Repeated Measures: Exact Methods in
MANOVA and Mixed Models

WEISBERG · Applied Linear Regression, *Third Edition*

WELSH · Aspects of Statistical Inference

WESTFALL and YOUNG · Resampling-Based Multiple Testing: Examples and
Methods for p-Value Adjustment

WHITTAKER · Graphical Models in Applied Multivariate Statistics

WINKER · Optimization Heuristics in Economics: Applications of Threshold Accepting

WONNACOTT and WONNACOTT · Econometrics, *Second Edition*

WOODING · Planning Pharmaceutical Clinical Trials: Basic Statistical Principles

WOODWORTH · Biostatistics: A Bayesian Introduction

WOOLSON and CLARKE · Statistical Methods for the Analysis of Biomedical Data, *Second Edition*

WU and HAMADA · Experiments: Planning, Analysis, and Parameter Design Optimization

YANG · The Construction Theory of Denumerable Markov Processes

* ZELLNER · An Introduction to Bayesian Inference in Econometrics

ZHOU, OBUCHOWSKI, and MCCLISH · Statistical Methods in Diagnostic Medicine

*Now available in a lower priced paperback edition in the Wiley Classics Library.

RETURN TO:
MATHEMATICS–STATISTICS LIBRARY
100 Evans Hall   510-642-3381

LOAN PERIOD ONE MONTH 1	2	3
4	5	6

All books may be recalled. Return to desk from which borrowed.
To renew online, type "inv" and patron ID on any GLADIS screen.

DUE AS STAMPED BELOW

FORM NO. DD 3
2M   11-05

UNIVERSITY OF CALIFORNIA, BERKELEY
Berkeley, California 94720–6000